WHY BITCOIN?

Published in the United States, Canada, and the United Kingdom by EnCompass Editions, an imprint and division of Compass Atlantic Inc. Canada.

Paperback ISBN: 978-1-927664-20-9

Printed in the U.S.A.

encompass
EDITIONS

WHY BITCOIN?

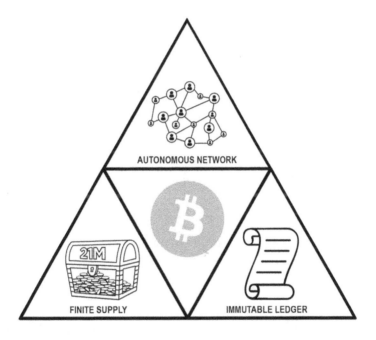

AUTONOMOUS NETWORK

FINITE SUPPLY

IMMUTABLE LEDGER

THREE TRANSFORMATIVE SUCCESSES

Brooke C. Williams

To Natalie and Owen, who taught me how to learn again

CONTENTS

LIST OF ILLUSTRATIONS

INTRODUCTION

Nearly fifteen years ago, we were given a gift.

Bitcoin's[1] creator posted some thirty thousand lines of computer code to a public bulletin board. We thought we knew what it was: a form of internet money that might or might not become useful. The code slowly and autonomously percolated from its initial, more technically-minded supporters to a global cadre of computers and bitcoiners that make up the largest decentralized "digihuman" network on earth. Much has happened in Bitcoin's history, and the combination of more minds considering its possibilities, the network's reactions to extraneous events, and its own growth and maturation have led us to understand that this wasn't just some gift of a money system (as important as that may be). This gift may also be a governance system, a verification system, an economic ideology, a global political force, a value storage and transfer facility, and a permanent record retention system. We have discovered that it may have many uses.

There is a reason why over these years, some of our smartest humans have spent a lot of time wrangling over this thing called Bitcoin. It's excitingly complicated. It also involves math and money and computers—things that often make peoples' heads hurt—so it's going to take some understand in places where people don't perceive an immediate need for change.

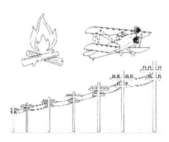

Illustration 1: Transformational Technologies

Most of the transformational technologies we have invented have been pretty explainable in simple terms. We learned to use fire, perhaps by watching what wildfires do. We harnessed the power of electricity in wires; you can see it power machines. We moved things fast enough to lift off the ground and fly like birds. We send information over wires all over the earth because we connected all our computers and phones to the same network. The innovative part of these advances is easy to see and describe; they are tangible.

This new creation is a little different. As we move further into the electronic age, more and more things are possible in intangible form. Without needing to be grounded in physics, our imaginations have been free to create goods and services without the limits with which our history has been forced to work.

This digital canvas has allowed an artist, Satoshi Nakamoto, to weave together three transformative ideas, along with numerous clever sub-elements, into a tapestry of innovation. The new technology is open-ended, as well as being open-sourced, meaning that it will be used and modified ad infinitum.

This new element of being intangible, of not needing to heed the constraints of the physical world, also means that we cannot know now what might be wrought in the future with such a tool. A digital tool that acts like a "typewriter" created out of code may be modified into a different tool that acts like a "microscope" by an enterprising developer, creating an entirely new domain to explore. You can't 'morph' a physical hammer into anything else, at least not easily.

A link to the physical world is not needed for innovation and creation. It is extremely important in the assessment of value. In a digital world where duplication is easy, there needs to be some way to verify validity and ensure scarcity for items we want to be able to hold value, or to use as mediums of exchange. As we will see, the investment of energy is a feature critical to Bitcoin's security not only in terms of defending against attacks but in the very assignment of scarcity and corresponding economic value to an intangible asset.

Satoshi was intent on creating an uncontrolled currency that would serve the people. Although I describe it here in terms of concepts, what Bitcoin does in practice is straightforward: process valid transactions submitted by unrelated nodes and record the answers onto a ledger that all nodes maintain. This allows for secure transactions of value to occur on the internet without needing to answer to any authority. No corporate intermediaries, no banks, and no governments are needed. Of course, these existing traditional entities can choose to restrict onramps to the new network, which makes it a longer, slower process to adopt, at least for some use cases.

Illustration 2: Intangibility

The variety of topics that Satoshi needed to address in order to perfect his version of an electronic currency has left us busy trying to figure out how many different ways the technology can be successfully adopted. There is no question that his primary product —bitcoin the currency— has done and continues to do exactly what was intended, without interruption or subjugation. It seems like we will be developing legitimate use cases and platforms for this technology—there are certainly thousands of developers at work doing just this. My take is that appropriate iterations of the technology must follow Michael Saylor's three requirements (as later detailed): that it be ethically, economically, and technically sound. While that seems a straightforward task, I am not sure that any other blockchains have been able to pull it off yet. To be as ethically sound as Bitcoin, a blockchain would need to be absolutely transparent, permissionless and fair, and totally decentralized—qualities which I do not believe exist in any other platforms at the moment. They will come.

Given the environment of online security, it is truly impressive that the Bitcoin network has not been compromised in its fifteen-year history. Every last hack or theft about which you have read have occurred *outside* the Bitcoin network, mostly in centralized platforms. Lost coins are not the fault of the protocol, but of careless humans. Not only has Bitcoin's security not been breached, it has been fully operational for 99.98% of its existence.[2] For this reason, I choose to call Bitcoin and its components successes. Of course, nothing is forever, but what we have come to understand is that Bitcoin's network appears to be *antifragile*. Attempts to take over the network actually make it stronger because in order to try to get control you must first contribute to its strength.

There are already a number of excellent books that describe the phenomenon of bitcoin from important angles, such as logistical tutorials on ownership and wallets, historical accounts of the disruption caused by its permissionless and libertarian roots, and prognostications of how bitcoin will change our systems.

This book is not a primer on how Bitcoin functions or how to use it. It has turned out to be a tribute to all the various components of the Bitcoin code, within which I identified three as historic firsts. Anyone exposed to the Bitcoin world can attest that there are a great many threads to follow in the process of trying to understand a multipurposed invention. While I have enjoyed digging through the research and learning from many minds, the purpose of this book is to try to focus on the intellectual achievements—the concepts, irrespective of actual use cases, that this new technology represents.

It's easy to start a list of neat things that Bitcoin does and the elements assembled to do these things (I put my list at the end of this book; you can make one, too). What I am trying to show in this writing is that while some parts of the platform are upgrades to what we already do, such as a better form of money, a superior verification system, and a quicker and easier way to send payments etc., there are aspects of Bitcoin that are *transformational* in that they have never been done before and that, correspondingly, we don't know exactly how far their use cases can be extended.

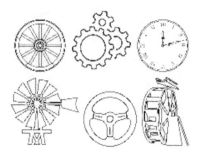

The Wheel: A Transformational Technology

Illustration 3: The Wheel

Transformational technologies can be defined as innovations that are open-ended in their potential use cases. Unlike a hammer, which has truly one use, a transformational tool is one that has been invented and proven efficient in a great many uses, most unimagined at the time of creation. The wheel, for example, has made itself useful in a number of sectors for many different uses, from transportation to timekeeping to plumbing.

The primary feature of a transformational technology is that the *springing intellectual innovation* is unique—humans have never before successfully employed such a tool.

Bitcoin does three transformational things—*things that have never been done before*:

1. Employs a global, entirely decentralized and **Autonomous Network**
2. Generates a **Finite** and Perfectly Inelastic **Supply**
3. Creates a Visible, Permanent and **Immutable Ledger**, or record

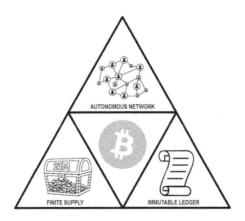

Illustration 4: The Triangle of Innovation

We don't know at this point what other developments Bitcoin might spawn as use cases or further tools, but this threesome of concepts is a starting point to allow us to develop new, cooperative platforms and protocols.

I have based this book's structure on the idea that these three successes are the linchpins to understanding Bitcoin's purpose: they are *transformative*. Satoshi's other clever ideas and innovations I call *disruptive or innovative*, elements of which are peppered into the three transformative successes. For example, cryptography is integrated into the management of the autonomous network and into the protection of the finite supply. Many of these subject areas are disruptive in their particular applications, such as cross-border payment functionality, but I see them more as innovations to existing ideas rather than transformational and original. By no means is my list complete or completed; these are the topics that most caught my attention.

I should also clarify that, while some might consider the internet itself to be the first autonomous network (h/t Jeremiah Soucie), I see the TCP/IP protocols as more a part of the underlying infrastructure. All networked computers require electricity, hardware and software, and an internet connection/protocol. Like any other network, Bitcoin requires the internet to function. To me, this makes the internet a different animal—a medium without the specific intentions of a network.

I also do not seek to minimize the achievements of some of the early decentralized networks like Tor and BitTorrent. These networks enabled file sharing without centralized control and can be considered autonomous and permissionless. As far as my research goes they have not, however, been able to achieve usage of more than perhaps 25 million users (less than half of 1% of cellphone users worldwide). They are honorable precursors to the technological breakthrough of Bitcoin.

There is a lot of overlap between the disruptive and the transformational ideas that are woven into the technology that is Bitcoin. As the adoption and usage grows, we will be adding to the list of new ideas birthed from this exciting new technology. Part of the difficulty of explaining bitcoin to newbies is this "problem" of so many flavors. A serious chocoholic may never explore most of the other 31 Baskin Robbins flavors; to her, ice cream may mean cold chocolate. She may never appreciate pineapple, but I maintain that she can learn to appreciate the people who go to BR31 only for the pineapple, just as she goes for the chocolate. The beauty of bitcoin is that it offers many flavors to many people, and its culture of innovation and acceptance says, "That's OK!". The goal of this book is to expose many people to many flavors and let participation and innovation flow where they may.

When Bitcoin naysayers assert that "there is nothing backing bitcoin", the argument is usually framed within the critic's perspective that the technology has not produced anything useful or that it may produce things that challenge existing privileges. This leads detractors to brand current use cases as useless, criminal, or out of control—ideas that actually contradict themselves. These criticisms ignore these three **intellectual innovations** brought to us by the inventor of the code and the further collaborative work of talented developers. It is becoming more difficult to use the value argument against bitcoin as even Larry Fink of Blackrock, the world's largest asset manager, has just acknowledged that bitcoin is an 'international asset". When existing traditional firms like Blackrock, Fidelity, et al. bless bitcoin as they have recently done, the quantity and quality of people joining the movement begins to drown out the uninformed naysayers.

At this point in 2023, however, we have also seen the downside of innovation with the collapse of the market due in large part because of chicanery and malinvestment. We are in the midst of another cyclical bear market, compounded by macroeconomic events in traditional finance markets and the overall economy. Many have lost the thread of the Bitcoin story among the numerous 'crypto' memes. We have ridden waves of successes and failures over the last five years, epitomizing the fear and greed of human endeavors.

Illustration 5: Interconnected World

The financial systems that have been put in place for the last 75 years or so have revealed weaknesses from time to time through failures of banks, governments, sectors or agencies. We have paved over the basic rails with so many layers of physical, digital, leveraged, electronic, printed, securitized, structured, or rehypothecated forms of money and credit that nothing is simple anymore. If you are proposing to change the base layer of all these measures of value (imagine the audacity!), you will clearly face considerable resistance, and it will take quite some time for changes to roll out through the system and the inevitable bugs to be repaired. The sheer scale of such an endeavor means that billions of people will be involved; there needs to be some sort of decentralized cooperation involved.

One can make the same case for other control mechanisms in our societies—they become layered and complex, with many parties being interested in preserving the status quo. That is why tinkering with a new technology in areas such as finance will generate considerable stress— perhaps more so than with an invention that is simpler and obviously better than its predecessors, such as a smart phone.

Blockchain technology is essentially a new database organization system, directed by a threesome of innovations blended smoothly and delivered to us in one package in 2009. Using Bitcoin as a template, other chains have and will be created that may be able to explore exciting new uses while maintaining the strict code-is-law approach that guides Bitcoin. With that in mind, at this point in time, it seems to me that bitcoin is the only truly decentralized token of any substance.

It is a pretty tall order to duplicate Bitcoin's rather mysterious origins and bootstrapping evolution to a store of value. We must identify a suitable mechanism to allow developers to move from a centralized 'beta project' status to a legal, decentralized network. As autonomous networks are not yet self-creating, there must be a legal and regulatory framework for developers to be able shepherd early supervision of a project into a mature, self-directed network.

If we are serious about the decentralization and autonomy envisioned with Web3, a protocol must be as permissionless and open as Bitcoin and without elected, appointed, or selected cohorts as overseers. If it comes to pass that some of the other existing tokens are judged to be securities and sued by the US Government, the very fact that there is someone to sue indicates to me a lack of sufficient decentralized and distributed control over the chain.

We do need to admit that the idea of *everything* 'going digital' should not be a terribly foreign one given that we have taken great pleasure in moving to electronic/digital so many other things in our lives from audio/video to phones to workstations, email and social media work product, and even our automobiles. Why would we not eventually just digitize everything of value?

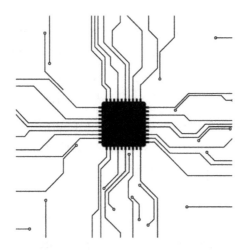

Illustration 6: The Microchip

PART I

A GLOBAL DECENTRALIZED AUTONOMOUS NETWORK

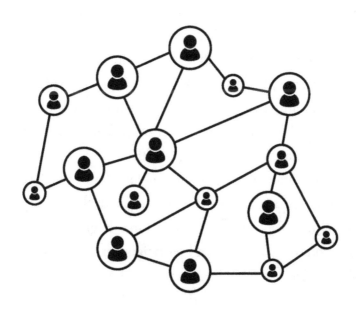

Autonomous Network

- Verification
- Money and other use cases
- Absence of an issuer and/or controlling party
- Existing tokens
- Business tokens
- Tokens of value
- Currency vs Money
- Consensus Protocols
- Trust the system, not the node

Bitcoin is the world's first completely decentralized and autonomous computer network. It is unique not only because it is the first, but also because it is pointless to duplicate it exactly, as it already exists and has much capacity. It's like proposing a second internet. Critics may retort that we already have a financial value system—true enough. But Bitcoiners, among others, have had the patience to examine the weaknesses and unfairness of how things work and feel that there are many things that could be improved (that's why we have Bitcoin to begin with!)

Every now and then, a new technology helps displace the way 'things were always done.' Bitcoin reached this position in an entirely organic way, slowly gaining fans. It grew over time with increased voluntary interest and participation. Its inventor(s) solved multiple problems to make this happen. The introduction of a faceless, leaderless protocol by a still-unidentified creator is worthy of science fiction, yet it may prove to be a tool for survival over the next several centuries.

This section is dedicated to the interesting concepts in Bitcoin that are associated with its ability to operate such an independent network and what this means to future systems.

THE ROLE OF VERIFICATION

The rise of Bitcoin has given humans a new tool for an age-old process. The tool is a decentralized, peer-to-peer, permissionless and transparent protocol. The age-old process is the process of verification.

> # Ver·i·fi·ca·tion
> *noun*
> 1. The process of establishing the truth, accuracy, or validity of something.

Let me explain.

Human beings are undoubtedly social creatures. According to a 2009 study by the World Bank, 95% of the earth's population is concentrated in just 10% of the land surface. We tend to live together, as our cities and towns demonstrate.

We reap great rewards from this social environment. The exchange of information due to our overlapping and interconnected lives rises far above the level of other social creatures such as ants or bees. The cooperation among these hive animals allows them to more efficiently fill their ecological niche, sharing in such needs as shelter, food, and defense. By virtue of our bigger brains, humans have extended the benefits of communal living to improve many, many other aspects of our lives.

The benefit of having millions of ideas and inspirations is indisputable; it is the source of much of our technology and the improvements to our daily lives. However, this explosion of ideas must be—and always has been—balanced by the equally powerful realities of feasibility and moderation. Not every idea works. We need to be able to efficiently discard the unprofitable, the inefficient, and the unworkable so that we can make better use of true innovations.

The process of *verification* has been our timeless and universal method to separate the useful from the useless. From the time we are young children, we learn that we must employ some form of proof, of double-check, of *verification* for our ideas. We start out by asking our parents. We then move on to friends, contemporaries, teachers, mentors, professors, instructors, experts, coworkers, bosses and others we trust. We learn to *verify* on our own by using reference materials—studies, books, papers, experiments, histories, the media and other records and sources of information to confirm or refute our thoughts.

Throughout our history, sources of verification have generally been limited to the people and the proofs that we *trust*, know, or that have some sort of influence over our lives. My aunt is a good source of chicken recipes, *I've used them.* A policeman can check that traffic is flowing at an acceptable speed, *he can penalize me.* Whether direct or indirect, our sources of verification are usually known to us in some way. How could we possibly trust an unknown verification source?

In academic and scientific communities, we have developed *verification* systems of reference and citations that bring credibility to newer ideas. Using prior proofs or results is a way to streamline our knowledge gathering— we don't need to reinspect every single component of a new math problem or a scientific experiment when the building blocks for which have been verified and reverified countless times by independent people or entities. We rest on a history of prior trusted verifications.

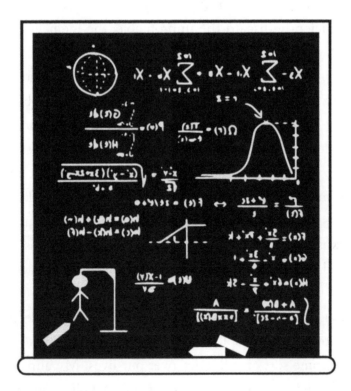

Illustration 8: Academic & Scientific Formulae

Throughout our history, we have relied on verifications made by:

1. close and trusted personal references
2. authorities with leverage over our lives who command that we believe in an idea
3. established experts who have credentials we believe, or
4. a documented history by named individuals of repeatedly proven 'facts' in a subject

Until now.

For the first time in human history, Bitcoin has given us a way to conduct *verification* that does not rely on trust in specific people

---

Although we have relied on these four traditional methods of verification for millennia, there are some clear drawbacks to trusting specific people:

1. Not everyone is trustworthy
2. Authorities are not always correct and do not always act beneficially to their subjects
3. Even experts can make mistakes
4. Scientific and academic references often rest on assumptions which are later proven false

As examples, I can cite Bernie Madoff or Sam Bankman-Fried. I can refer you to various cults or religions in our past with credos or belief systems that were not appropriately verified, but were nevertheless practiced by their peoples. I can mention how the world before the scientific method was practiced relied on simple observations that did not survive subsequent discoveries. I can remind you of the European scholars who were convinced Columbus would sail off the edge of the world. We can talk about how charismatic leaders like Alexander the Great, or Napoleon, or Gandhi, or Mao Zedong, or Stalin, or Franklin Delano Roosevelt have convinced their peoples of the legitimacy of their motives and doctrines, irrespective of ideology.

Most recently, the James Webb Space Telescope has demonstrated that the Big Bang may have just been another theory based on incorrect assumptions from incomplete observations of our universe. After 100 years, we might have to rewrite some of our understanding of our origins.

When our emotions are involved, we stand even less of a chance of conducting clear-eyed and accurate *verifications*. Today's divided political climate in the US has shown us that it is not hard to sway people away from using proven and accurate sources of verification and towards emotionally-pleasing but often incorrect verifiers. Those people who were afraid to be left out of a housing boom in 2005, or those who got an expensive college degree in the last 20 years, were following "conventional wisdom" that turned out not to be so wise for many. Nearly every instance of human suffering I can think of can be linked in some fashion to this faulty mechanism we have been using for *verifying information*. As it turns out, we are wrong or misled... a lot.

Illustration 9: Leaders as Verifiers

What if we were given a new way that we could exercise optimum validation and verification by employing a large cohort of users sharing an identical computer code and using mathematics, algorithms and cryptography? A network that could not be "wrong" in individual human ways, that would be highly transparent, permissionless, and which would create an immutable record of all its activity. That sounds pretty cool, almost too good to be true, and certainly worthy of the 'immaculate conception' legend surrounding its introduction.

Of course, the cadres of "trusted intermediaries" we have used for centuries will have something to say about decentralized governance. Politicians, banks, social leaders, experts, academics, and anyone used to wielding authority will probably not like the idea of the 'wisdom of the crowds'. Many of them will feel a need to "protect" us from ourselves, even though we are collectively voting to use this tool and are about to breach the one millionth bitcoin wallet holding at least one coin. We can expect continued resistance from all quarters.

A **new system for cooperative verification** is at the essence of the Bitcoin revolution.

Any new technology must go through testing and verification periods of its own. One way to accomplish this is to release it to the public. Today's connected public is arguably the toughest testing ground for any new product—there is no lack of supporters, testers, reviewers, doubters, bad actors and other individuals who are very willing to test the boundaries and capabilities of a new tool.

Throughout our history, creators of innovations have often kept their recipes "secret". This is obviously done to retain control and ownership over the invention, not to mention potentially making a lot of money.

The first mentions of Bitcoin were published on internet chat boards frequented by the Cypherpunks, a libertarian/anarchic collective of computer programmers. Part of the Cypherpunk ideology is to wrest control of the world from super-controlling representative governments and return power to the people. Apparently, in this vein, the Bitcoin code was publicly and anonymously posted online, giving access to the technology to anyone who wanted to copy it.

Very different from the 'walled gardens' proprietary platforms and algorithms of Web2, the Open-Source model encourages open collaboration and free distribution. This is part of Bitcoin's success: a permissionless, decentralized network that offers an alternative to the traditional banking system (and other information networks) in the hands of many unrelated nodes that just can't be stopped.

Illustration 10: Decentralization as Defense

After surviving a decade and a half of exposure to the public without being destroyed, it seems reasonable to think that this new protocol for verification, Bitcoin, is indeed sound.

Michael Saylor, chairman of MicroStrategy, Inc, one of the largest corporate holders of bitcoin, analyzes the successes of this period of public ownership and awareness as follows:

1. Bitcoin is **ethically** sound. It is a commodity without an issuer. No person or group profited excessively from its creation or controls it now.
2. Bitcoin is **technically** sound. It has not been 'hacked', grows stronger each day, and it has operated continuously for 99.98% of the time for 14 years.
3. Bitcoin is **economically** sound. It creates a positive, market-based value proposition for miners and token holders alike.

While easy to understand, Saylor's assertions were anything but easy to accomplish. The code had to account for challenges in many areas, from computing expertise to network dynamics, to human behavior. There are many alternative cryptocurrencies, developed from modifications to Bitcoin's available code, but none has enjoyed the success of the original.

So, despite many calls for the death of Bitcoin over the years and attempts to change the core code itself, this new verification tool has persisted and flourished—indeed the network now enjoys a market cap greater than America's largest bank, JP Morgan Chase. That figure just reflects the price of the tokens, not the value of assets employed on the network or the transactions processed.

On May 22, 2010 "**Pizza Day**", 10,000 bitcoin were paid for two large pizzas in the now-famous first known bitcoin transaction for real world value. Quite naturally, as soon as bitcoin gained economic value, the doors were opened to anyone interested in creating non-sovereign alternative networks of value.

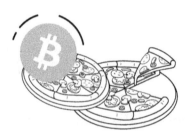

Illustration 11: Pepperoni bitcoin

Most of the alternative blockchains since created have failed to meet Saylor's criteria. It turns out that being ethically sound is a particularly difficult challenge for humans—there is no shortage of people who would like to gain fame and fortune, even at the expense of others. It helps us understand just how unique and special Bitcoin's birth story is and how the legend of Satoshi Nakamoto, aside from being a wonderful mystery, may have been one of the only ways that such a new protocol could have been released, survived a distributed existence, and eventually flourished.

Some of the early blockchain developers realized that, in addition to creating new wrinkles to the bitcoin blockchain standard, the technology might be modified more significantly to allow for verification of other elements critical to our lives. Money might be the initial and primary use case, but there may be others that end up just as noteworthy:

- Identity
- Ownership of assets (and liabilities)
- Information
- Processes—business product lines, credentials, certifications, approvals, auditing
- Safety—food, products, medicines, transportation, building, vehicles

...and more.

MONEY AND OTHER USE CASES

Although the stated reason for the creation of bitcoin was as a "peer-to-peer electronic cash system", a critical component of this system was the ability to verify the ledger, or accounting, for this electronic cash. Many books on bitcoin focus on its role as a money. This entails learning a good deal more about money than the 'average joe' starts with. While I do not want to get sidetracked in these discussions, learning the basics does help many understand some of the capabilities and characteristics of such a digital asset, so we will briefly cover them here.

Prior efforts to create digital money encountered some of the verification problems specific to currencies and networks such as the Double Spend Problem and the Byzantine Generals Problem. These relate to being sure that the money is not somehow copied or duplicated and that false transactions submitted to the network are identified and discarded.

In solving these problems, Satoshi also gave us a template that may be overlaid on other use cases that might not be quite as complex as a digital currency.

Money has some well-known roles and characteristics that make it especially challenging to manage—more so than the other use cases for verification I listed above. Ownership or identity verification, for example, is a more static process with many fewer characteristics. A blockchain can easily handle the ledger requirements of recording house titles, for example. It seems only a matter of time before ways are found to employ blockchains in these other less complex use cases.

Money has existed for thousands of years as representative of value. The "When?" question of money (as in "When did we start using money?") came up pretty quickly when people realized that representations can be far more efficient than actual goods...which also answers the "Where?" and "Who?" questions of money—wherever there were civilizations active in commerce and trade.

Most people are intuitively familiar with two basic roles of money: a Store of Value and a Medium of Exchange.

A Store of Value: I have a way of preserving the value I earned today, and did not spend, until tomorrow when I might want to spend it. When I am paid for my work in dollars, I can keep them until I am ready to spend them. We also call this "Saving", and savings are easier to keep when they are preserved in an agreed and accepted form, like money. Gold is a traditional store of value.

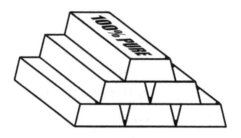

Illustration 12: Gold Bars

A Medium of Exchange: A big reason I am willing to keep these dollars into the future to spend later is because I know they'll be widely accepted by others in trade for the things I will want then. My dollars act as a common go-between, or *medium,* for the work I did already and the things on which I want to spend my savings.

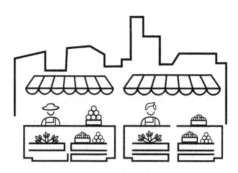

Illustration 13: Merchants

Having an acceptable medium of exchange solves an age-old problem with barter known as the *double coincidence of wants.* Why would I trade anything with you if you don't have anything I want? A free-flowing medium of money greatly increases the number of feasible transactions.

Critically, there needs to be enough adoption of a money for it to be considered an acceptable medium (that's the 'widely accepted' part). Greater numbers of users and a higher market cap (total value in circulation) also help ease the volatility a young currency may experience as large players can influence the trading marketplace—stability is preferred.

One can see that if I am going to place so much faith in something that I'll need to use for my future survival, *verification of its ability to do so is very important.*

These two roles are the reason why every money all over the world has come into existence, and so believing in its veracity is the whole point. We can think of these as answering the "Why?" question of money. I also like to think of the two uses in terms of time use. I want a medium of exchange available immediately and easily so I can buy a cup of coffee. I want a store of value to be available to me a long time from now, hopefully with an even greater purchasing power.

Springing from these two essential uses come some easy-to-understand characteristics of things that are suitable to act as money.

These are not difficult concepts. If you just think about what kind of guarantees you would want to have in order to preserve the value you earned through your labor, you could have come up with these on your own:

1. **Durability**. It's why we don't use bananas as money. No guarantee that it will be around when I need it.

2. **Portability**. It's why we don't use elephants as money. I need to be able to easily move my medium of exchange around—to the market and home again.

3. **Divisibility**. Another reason why we don't use elephants as money. It's hard to trade a quarter elephant. We need a good that can be fractionalized or is small enough in value to deal with the smallest transactions.

4. **Scalability**. Yet another reason why we don't use elephants as money. Along with the ability to handle small transactions, a money needs to be plentiful enough to be owned by all the users who desire it. (There aren't enough elephants for everyone!)

5. **Identifiability**. This is why much of our modern money is branded with our leaders' images. In order to be accepted, the money needs to be readily identifiable as the value it represents. (Elephants might actually work for this characteristic!)

6. **Fungibility.** Together with Identifiability, this means that the tokens of value are interchangeable—any dollar is the same as any other dollar even though they are not exactly identical (serial number, signature, etc.) and therefore, they are each reliable as individual stores of value and suitable for exchange.

Taken together, a successful store of value and medium of exchange that can be used as money needs to be durable, portable, divisible, scalable, identifiable and fungible. Notice that the first four characteristics have to do with money's physical reality—they answer the "What?" question of money ,as in "what can we use as money?"

The last two characteristics have to do more with usage—how humans go about assigning value to a representative token (more about the word "token" in a minute). These answer the "How?" question of money and are the elements that require more verification.

The final characteristic of a good that can be used as a money is **scarcity.** Yes, although this seems a bit counter to the scalability characteristic mentioned earlier ("there must be enough"), scarcity is perhaps the most important of the characteristics in that it defines the *value* imbued in a money. In order to carry value, a money must be relatively difficult to acquire. It's why we don't use leaves as money.

Obviously, there will need to be a *balance* between **scalability** and **scarcity** that both allows for sufficient quantities for medium of exchange uses without sacrificing the value of each token by not having some sort of limiting function on supply. This has been the tricky part. Inevitably, civilizations have improved on the scalability of money but failed to control its scarcity.

Our Achilles' Heel

There have been thousands of versions of money created in our history. Every one of them that failed in the long run (as far as we know) has collapsed because of a problem with scarcity. Actually, a lack of scarcity. Every time politicians and leaders begin spending government money or public resources, the same pattern is repeated. They spend too much. Then, to address a budget shortfall, the leaders create more money. Sound familiar? The most frequent reason for over-spending? War.

War has a way of physically destroying the losers and financially destroying the victors.

Obviously if a new money is not somehow linked to some cost, if it is too easy to create, the new money just devalues the existing money supply until...it's worthless.

The other way money has suffered from oversupply is via the emergence of a new technology (or the arrival of competing humans), making the creation of more money easier and again leading to an unsustainable economic outcome.

Scarcity has always mattered.

ISSUED VS ISSUER-LESS

In Bitcoin, the computer code controls everything, including the issuance of tokens. Because of this, bitcoin behaves more like a modern version of a commodity-money, like gold coins than it does a fiat currency, like dollars.

Commodity-based moneys are generally not created (although they may be grown, refined or processed) by any particular persons or entities; they are issuer-less. Some like to say that Mother Nature is the issuer. Precious metals, crops, shells, furs and other items historically used as commodity-moneys come from nature and are available to those who wish to collect, refine, or grow them. They are not "issued" by a particular group, they are not "restricted" by humans, and their supply is not limited. Assets like these, without an issuer, are called commodities. Commodities also cannot be trademarked or claimed by decree. They belong to their bearers.

> ## Com·mod·i·ty
> *noun*
>
> 1. a raw material or primary agricultural
>   product that can be bought and sold,
>   such as copper or coffee.

As bitcoin are brought into creation by the miners who conduct the cooperative verification, there is no singular source and no issuing party. Like gold, bitcoin comes from a work process applied to an environment. The distributed computer code provides an unchanging environment, mimicking the role of Mother Nature. New bitcoins are issued by the operation of the code.

I used the words "generally not created" above because there have been efforts over time to influence use of commodity-moneys by making them identifiable in specific ways—by creating gold coins with an emperor's image, for example. Arguably, though, the images did not change the value of the tokens used but did assist with the verification process in that they added to the perceived trustworthiness of the coin. A gold coin marked with an image known to represent some authority has more credibility than an unmarked lump of metal. Therefore, this characteristic is more important to the Medium of Exchange use of money than the Store of Value use, as its authenticity can be proven in other ways than by bearing an identifiable image.

Because of the transparency of the blockchain's transactions, every single fraction of bitcoin can be traced electronically from its creation as a block reward up to today. This characteristic makes bitcoin a "super-commodity" in a sense because no other commodity comes with a ledger of all prior owners and transactions throughout history. This transparency effectively makes bitcoin more 'provable' (or *identifiable*) than other commodities. Its lack of centralized control or an issuing authority also makes it more immune to the machinations and manipulations of humans.

Illustration 14: Transparent Security

If you have played the board game Monopoly, you know that a banker is selected from the players to be in charge of the bank. This assignment works, despite monetary incentives for the banker to cheat, because the operation is entirely transparent. Any cheating by the player-banker ought to be exposed relatively quickly because most players pay attention. This is similar to Bitcoin's model of transparency.

The idea of issuer vs issuer-less is also important in the later discussion on legal status of other blockchains and ledgers that have been created. Institutions, boards of directors, copyrights and venture capitalists are all signs that a particular digital asset is not issuer-less and indicates that control over some element(s) of the asset is in place. We have had many examples in the last few years as to what can happen to an issued crypto token—issuers make changes.

"A lot of people automatically dismiss e-currency as a lost cause because of all the companies that failed since the 1990s. I hope it's obvious it was only the centrally controlled nature of those systems that doomed them. I think this is the first time we're trying a decentralized, non-trust-based system."

– Satoshi Nakamoto

THE WORLD OF TOKENS

Over the last two years, the rise and fall of many 'crypto' projects has reawakened peoples' awareness of the word token. Although those of us born before the internet associate Monopoly board game pieces, city subways, or amusement park tickets with the word, tokens are part of the new, intangible, digital future of technology. They are not just tickets, worthless other than for carnival rides. Despite the black eye given to the word after the collapse of digital exchanges and fraudulent shenanigans in the last years, digital tokens offer a way for value or other elements of our culture requiring verification like ownership or identity to be represented and traded on a digital, intangible basis.

It seems pretty clear that this will be very useful in an internet-based world. Even sovereign nations are discussing creating CBDCs (Central Bank Digital Currencies)—government-managed digital tokens. This may be a double-edged sword as the ability of governments to track our data hands them potential control over our lives in undesirable ways.

Illustration 15: Game Tokens

I like to use a broad definition of the word token, which then includes any representation of our elements of culture that require verification or authentication. In fact, *we already use a great number of tokens* to simplify things like ownership, identity, rights etc.

The analog world is full of tokens too!

TOKEN	ISSUER	USE
Your Drivers License	DMV	Identity/Authorization
Your Passport	U.S. Govt agency	Identity/Nationality
Your Birth Certificate	County Records Office	Identity/Nationality
Your Social Security Card	U.S. Govt agency	Identity/Benefits
The Dollars you use	U.S. Govt agency	Medium of Exchange
Your House's Title	County Records Office	Ownership
Your Car registration	DMV	Ownership
Your credit cards	Banks	Monetize future earnings
Government Bonds you own	U.S. Govt agency	Store of Value/Income
Stocks, Bonds, Mutual Funds	Banks	Store of Value/Income

Illustration 16: Universe of Trust

All of these tokens are associated with a ledger that keeps track of the issuance. They all have issuers who exert control over their issuance, the ledger, transactions, security, the rules of use, and other aspects - they are not all representative of *tradeable* value. None of these tokens is decentralized, all operations are controlled by the central authority and the tokens must be used strictly in conjunction with the determination of the authority, not the crowd.

It makes sense to forecast that digital tokens – cheaper, faster, easier to use than real-world representations – will become our vehicles for many, many use cases.

My Silent Generation mother laments that she cannot use paper tickets for events – she now must figure out how to download the venue app on her phone to present he digital QR code ticket. Obviously the fact that she does not like it does not mean that digital e-tickets are not our future.

IDENTITY AND OWNERSHIP TOKENS

Bitcoin has proven that digital coins managed by blockchains can serve as tokens of value. Can digital assets also perform tasks in the areas of identity and ownership?

Based on the qualities of blockchains—cooperative verification and security through transparency—it would seem that public use cases such as for identity and ownership *may be just as appropriate* than for private transactions of value.

The very reason we need identity and ownership verifications is because we are a social animal and share many things. Since the time of private ownership, we have needed ways to distinguish between each other and our things. In the realm of value transfers, the verification process is focused on the value of the token. The taxi driver doesn't care who he drives or where they go as long as they pay the fare with an acceptable medium of exchange. When we start talking about rights, permissions, status and other social definitions, verification becomes important for reasons other than value. You don't want anyone else claiming your social security benefits. Your identity matters.

Until now, money has been chiefly a private tool, used to define interactions between individuals in the tribe. With the move to the digital world and the coming advent of CBDCs, we may see the involvement of our governments in previously private transactions to a far greater degree.

While the centralized control of currency by a government may not prove to be optimum to our freedoms, what does seem like a reasonable use case is to co-manage the things we already share, like the need to verify our unique selves and our possessions.

INFORMATIONAL AND PROCESS TOKENS

These types of tokens may soon be ubiquitous in the business world. Digital tokens are a simple tool to help contain verified information, such as in research, innovation and patents, copywriting, business data, and reference materials.

The business 'flow' of information through various departments or approvals can be more easily facilitated with verification tokens earned along the production path. The visibility of the process can be modified in private blockchains, allowing for confidentiality to be managed depending on users' needs and level of approval. Knowing that the information and data you are using have been verified by a 'swarm' of validators helps give credibility to assumptions, formulae, and presentations. Working with financial documents that have gone through sound verification processes can facilitate and speed up business dealings both in the ability to use more 'live' or timely information but also reduce or eliminate time-consuming auditing or delays on final settlements of traditional transactions.

Illustration 17: Smart Contracts

Eventually, we will get good at writing smart contracts—self-executing "apps" that live on blockchains and can receive and send data and payments coordinated in time. Once written, a smart contract can be hard to reverse, due to the immutability of the blockchain. This obviously makes it crucial that they are written without errors and bugs. Perhaps AI will step in to help us edit our smart contracts before they are recorded on the blockchain.

SAFETY TOKENS

A final example of a category of tokens is those used to indicate the validity, legality, or safety of a product, process, or service. We now rely on 'certifications' by industry groups that products actually are what they represent themselves to be. The sticker at the grocery store tells us that the apple was organically grown. There is little other way for us to be aware of the designation (or to even know what that means.) Here again, we are relying on a single 'issuer' of the designation and must trust its integrity. There may be ways— such as peer reviews in academics— that allow for industries to review products and processes on a supra-company basis and assign verification tokens, transparent and trackable for better consumer awareness and protection. Users of digital assets can be restricted from some uses by the community of users, much like bitcoin nodes can reject maliciously inaccurate transactions submitted to the network. Digital tokens allow us both rigidity of structure or protocol and flexibility of use or execution, all within a distributed, multiparty environment.

These logistical tokens may be useful for businesses and consumers, but the question is how to appropriately silo confidential or private information. It remains to be seen if there are uses for private blockchains that are superior to a centralized system for reasons we have not yet uncovered. In many instances, critics may be correct that a traditional centralized database may be more appropriate to business use. It is easier to see blockchains operating in more public arenas, such as local government oversight and identity- related data.

While tokens intended to keep us safer sounds like a good idea, there is always room for the misuse of good tools and Central Bank Digital Currencies (CBDCs) have the potential for governments to co-opt and abuse the technology. As has been widely discussed, a government can employ tokens for currency, identification, services, etc. but could tie their use to certain restrictions, standards, or conditions that subject the citizens to authoritarian control. Safer but also less free? It's all in the execution. The debate continues.

TOKENS OF VALUE: ASSET TOKENS VS LIABILITY TOKENS

Tokens of value have also been used as mediums of exchange for thousands of years. Bitcoiners Saifedean Ammous and Lyn Alden have independently written excellent analyses of the historic use of moneys, to which we refer the reader.

The first versions of tokens used by civilizations were collectibles that represented value to their users. Shells, tools, metals, furs, glass beads, and many other items were traded in marketplaces and villages worldwide. These tangible goods not only represented value, but that value was also inherently present in the materials or goods traded—the value was in the money. In other words, commodities themselves were the medium of exchange.

Legitimate commodity-moneys also shared the feature that they took work and effort to produce. Metals must be mined and refined, natural assets such as shells must be collected, fur-producing animals must be hunted, and their pelts processed. The mechanism for consensus identification of these moneys is usually visual: the "Proof of Work" is evident in that the commodities used are known to be rare or involve some sort of collection or manufacturing—processes that limit available supply and result in the desired characteristic of scarcity.

Commodity-moneys had some difficulties with some characteristics like identifiability and fungibility because things in the natural world are often not perfect— not perfectly alike, not perfectly formed, not consistent in content. There is also room for commodity-money to be modified in ways that are very difficult to verify. The content of a gold coin, for example, can be suspect. Beads, shells, furs, etc. are not identical and are subject to technological advances, such as the invention of machine-made versions or other artificial means of production that do not require the work needed to ensure scarcity.

Tracking Value : Tokens and Ledgers

Cowry Shells

African glass trade beads

Alexander the Great coin

Rao stones

Fiat currency notes: the US Dollar

Illustration 18: Tokens and Ledgers

This scarcity factor makes commodity-moneys intrinsically valuable. They take some work to get, and when in hand, they are bearer assets. There are no third parties involved in ownership; the bearer is the owner.

Commodities also have wide variations in their availabilities. Raw materials such as precious metals exist in "pre-determined" quantities. Rather than their availability depending on natural factors, our ability to locate and mine metals is the major driver of their supply.

Agricultural products and livestock, however, are much more available but dependent more on physical geography, climate, and more immediate conditions. So, even within the general category of commodity-based currencies, there can be considerable variation in the "hardness" or consistency of the value premium from one money to another.

Illustration 19: Gold Certification Seal

The rise of representative tokens came about with the trade, management and oversight of a commodity— tokens make transactions far easier. In a very natural way, the ownership of gold, expensive to safeguard and transport, morphed into gold IOUs or certificates of ownership, which were far more negotiable. Indeed, US banknotes used to have a gold certificate stamp, indicating the backing of the note by gold held in the US central banking system.

In 1933, FDR passed an executive order confiscating all gold from citizens and effectively disallowing its use as a store of value. Americans had little choice but to store value in dollars, eliminating the chance that gold could weaken what became the world's reserve currency.

The victory in World War II cemented the American dollar's role as the preeminent standard of value. Although we don't read much about it in our history books, the United States basically strong-armed the rest of the world into adopting its currency as the world's store of value. This wasn't particularly hard as the rest of the world was basically in ruins, and America remained essentially untouched, with a large and diverse economy, plenty of natural resources and available labor. American banks also held most of the world's gold reserves in their vaults, shipped there for safety from other nations at the onset of the war. The dollar was indeed "as good as gold".

Despite repeated requests, the US declined to return the gold to its allies, even 20 years later, when a French ship returned home empty, again rebuffed. In 1971, President Nixon oversaw a major change to the system: no longer would the US Government redeem dollars for gold. Severing the link to gold broke the restraint on new dollar creation, but it did ease the political friction in two ways: it ended American politicians' charade of needing to have gold on reserve to back dollar creation and lessened the foreign governments' pressure to return the bullion

At the same time, Saudi Arabia agreed (in exchange for aid) to denominate its oil sales in US Dollars, forcing all buyers to come to market with dollars in their pocket, an arrangement known as the 'petrodollar'. The petrodollar gave the dollar a natural and ongoing demand for dollars, which offset any concerns about currency supply. Energy, the driver of development, became priced only in dollars, and the other countries tied their currencies' prices ("pegged") to the US Dollar. It was no wonder, then, that the US was able to call the shots during the post-war period, and its expanding economy fueled a growing supply of dollars around the world.

In 1971, the link between tokens and their underlying assets was severed with Nixon's closing of the gold 'window'. No longer was the number of US dollars created limited to an equivalent value of gold held in the banking system. While such a decoupling had happened in history, never before had this happened to a nation with such a grasp on the global economy.

After 1971, the US Government has been able to create (and spend) any quantity of dollars its politicians have deemed necessary. Of course, one becomes a successful politician by pleasing constituencies—who wants to elect a Grinch? Spending more than you make is not a sustainable practice—unless you convince people that it is with arguments like its value is based on the "full faith and credit of the US government". So yes, dollars became a *faith-based medium of exchange.*

Removing a link to gold reserves was an absolutely critical step in our nation's past. The money we use now is very different from the money we used previously. Because it is not linked to a commodity or other valuable physical items, our politicians have been able to create as many dollars as they'd like without limit. Our currency changed there and then from being a bearer asset to becoming an obligation of the US government, a liability, with a third party.

This relatively new type of money is known as 'fiat' money. Fiat means "**by order or decree**."

...The next two charts show what happened to the supply of US Dollars when they became a fiat currency and the corresponding value of a dollar as more and more are created. *This is no secret, these charts come from the government.*

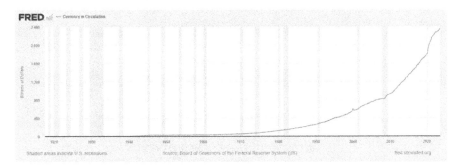

Illustration 20: Currency in Circulation

Most Americans are not aware of this modification of their currency from an **asset** to a **liability**. Most Americans in 1933 – and indeed today- did not understand what it might mean to have gold ownership nationalized. Once in control of the Store of Value asset behind the paper currency, there was no need to limit these tokens. The government controlled the assets and the ledger, acting as an issuer of a commodity. It took nearly 40 years for the US government to be forced to publicly recognize this control – and it was foreign government ledgers, not its citizens who forced the issue.

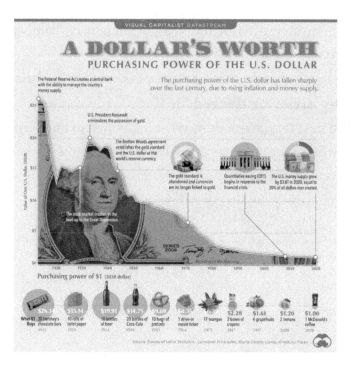

Illustration 21: A Dollar's Worth

An asset can be held free and clear by its owner and employed as he or she wishes. A liability—by definition—is a debt that comes with an attachment to another party and is subject to the conditions of the contract with that party and any changes that might be made by that party over time.

One of the reasons we may not really notice the cost of inflation is because in the US, with exceptions, we generally experience single-digit annual inflation. It's harder to notice when someone steals only 3% of your stuff in a year, but watering down your currency has precisely that effect on your purchasing power. Even our government's "target rate" of 2% monetary inflation results in a debasing of the dollar by halving its purchasing power in just 36 years.

Can you imagine living in a nation where it is obvious to you when you ask for an annual raise that what you get will not cover your loss of purchasing power in a year?

We are also not very clear on what the long-term ramifications of this change made by politicians, who are generally not known for their long-term thinking or their math skills, might mean for our future well-being.

As a very recent update, the US Senate is about to approve another extension to the debt limit, an allowance without any figures attached, letting politicians spend freely on top of the existing $31+ Trillion Debt. Twenty years ago, in 2003, the US Debt was $6+ Trillion, representing a 500% growth rate, or 25% per year, which even to a non-economist doesn't seem sustainable.

Real GDP, Real Wages and Trade Policies in the U.S. (1947-2014)
Index (1947=100)

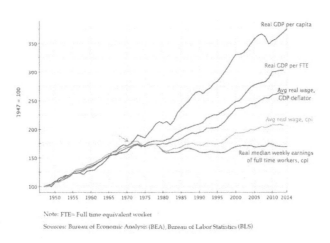

Note: FTE= Full time equivalent worker

Sources: Bureau of Economic Analysis (BEA), Bureau of Labor Statistics (BLS)

Illustration 22: Divergence: GDP vs Wages

A divergence clearly began in the early 70s. Were these intended consequences?

When you hear that bitcoin is not backed by anything, remember that bitcoin is an **asset** that is given value by the investment of energy into its creation, like gold. The US dollars in your pocket are **liabilities**—IOUs of the government—backed by faith in our politicians and their stewardship of our nation's budget.

One might hear the argument that any dollars created as a national currency have always been liabilities of the government, but when the dollar tokens were representative of specific assets (gold) their redeemability was 'guaranteed' at least to the degree they were considered 'safe'.

CURRENCY VS MONEY

Humans have been using various forms of tokens to represent value in the form of 'money', for thousands of years—it's nothing new. Another thing that should be pointed out to Americans as we use our dollars is that most dollars are already digital and do not "exist" in tangible form. If we use the word currency to describe the dollars you can keep in your pocket, it might surprise you to know that they represent somewhere between 5 to 10% of the total dollars in circulation.

Many people are surprised to find that most dollars, by far, are created by the banking system via loans made and *never exist in physical form*, only as digital entries on the lenders' ledgers. This has been made even more confusing by our government's response to the Global Financial Crisis of 2008 and the Pandemic Assistance of 2020, when we spoke about our government "printing money" to bail out banks and assist businesses and consumers. All the 'money printing' memes seem to indicate the production of actual currency that got distributed physically. This was not the case.

In fact, since most money is created in the private sector, by licensed banks lending money against the money they got from depositors, we are not really sure about how much money (vs currency) there actually is. If a bank owes you your deposits back on demand but at the same time has lent those dollars to someone else who is using them, the number of dollars effectively in circulation has been doubled. We have discovered in 2008 and in 2023 that this system of fractionalized reserves (lending out what you owe depositors) is fragile because we don't really have control over how many dollars are created. Banks are expected to watch their ratios and balance sheets and stay solvent because... who wants to fail?

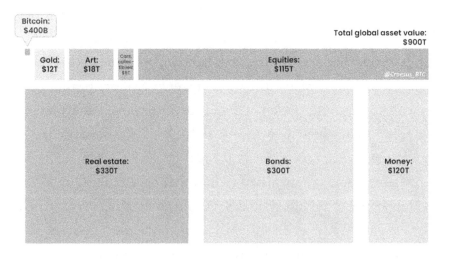

Illustration 23: Asset Classes

You probably have not been paid in cash, actual physical dollars, since you were babysitting or mowing lawns as a teenager.

Most of us get electronically transmitted paychecks, ledger transactions from our employers' banks to our banks. Similarly, we pay bills electronically or even by check, both of which do not involve actual physical dollars (*currency*). The youngsters these days seem to function without writing checks at all— nowadays, there's an app for that.

In the above illustration, I should point out that the box labeled "money" is not only physical money, as we have been discussing. Most of that "money" is the value of accounts kept on ledgers, like in your savings account at your local bank or like on the balance sheets of corporations.

So, not only have we been using tokens for thousands of years, we have been using *digital tokens of value* for decades...none of this is anything new! The other surprising consequence of the fact that most dollars are already digital and only exist as ledger entries is the idea that we really have no idea how many dollars there are. No one or no agency is keeping track of how many total dollars have been created in loans or in any dollar-denominated instrument or debt.

Dollars are not even supposed to leave the geographic United States, their residence as a Medium of Exchange and a Unit of Account. But while there are other currencies in other countries, many foreigners prefer to keep dollars as a Store of Value due to the weakness of their monetary controls. The Eurodollar is a nickname given to any dollar-denominated debt created overseas by any lending institution. We don't know how many Eurodollars are in existence because there is no institution to track them, and they can be created anywhere. I can buy a home in Europe and borrow Euros to make the purchase. If the lender requires that I repay the loan denominated in US dollars, they have created a dollar liability on my books. I must come up with the dollars to make the payments, so this "Eurodollar" debt is going to be repaid in the future with dollars out of the ledger-created supply. Because all debts are not disclosed to governments worldwide, we have no way of knowing what is out there...

The management of the world's money supplies has been the responsibility of sovereign governments for some time. Although there have been some examples of private moneys through history, over time, they have all succumbed to the control of governments. In order to enforce its recognition, users need to have confidence in (and go to market with) a money in sufficient volume so as to make it a useful Medium of Exchange. The confidence in a money is, therefore, in part due to its identifiability and usage, but also stems from perceived political durability. In other words, the ability of an issuer to 'guarantee' the durability of a currency can be considered as a part of its value. People will not want to use a money that might not survive.

In this vein, political and military power has played a role in the value and use of moneys, and generally the strongest nations have had the strongest currencies. This is another variant in the role of trust in our financial dealings. When we say that the US dollar is backed by the "full faith and credit" of the government, we really mean the full military and political might that the strongest world nation can bring to bear on others. This has also been called 'Proof of War'.

How, then, can Bitcoin compete with this political/military support enjoyed by the fiat currencies of sovereign governments?

The answer is that Bitcoin does not try to compete with nation-states in this area, but manages to generate its own version of political support essentially by pointing out the weaknesses in the existing sovereign systems. Bitcoin makes its own novel case by addressing the shortcomings of the system we have used for over a century, while enforcing a permissionless recruitment structure.

The single most common reason for the failure of state-sponsored moneys is the inability of governments to prevent an eventual inflation of supply. With its predefined supply schedule (and capped total supply) Bitcoin removes the role of an issuer and sets itself up to compete with fiat currencies in their Achilles Heel of distribution control. Although this means that the protocol must struggle to educate people to understand they can 'trust the code' as there are no issuer enforcement mechanisms, it also opens up the entire world as potential users of an alternative trust system.

Bitcoin has no military support whatsoever. This might be why so many intellectuals are attracted to its consensus mechanism: it is taking over by force of thought, not of physical might. In a David and Goliath matchup, it is the brains that survive over brawn—at least in the long run.

Illustration 24: Consensus Protocols

The premise of Bitcoin is that clear incentives and pitfall avoidance is a better growth mechanism than the tip of a spear. The genius behind Bitcoin is how well the consensus mechanism elements fit together such that its tokens of value are starting to take market share from the physical control-based sovereign coins. At the center of the incentives lies the truth that Bitcoin has a perfectly inelastic supply curve, which will not be altered.

CONSENSUS PROTOCOLS

While this book contemplates three historic "firsts" for humanity, we are also examining a number of disruptive innovations within these transformations that help to operate the network smoothly. The *Difficulty Adjustment, Heaviest Chain Rule,* and *Proof of Work* are three computations that nodes on the network follow to carry out the work of transaction verification and blockchain construction. These computations comprise the **Nakamoto Consensus**, the governing process for the blockchain.

While the Proof of Work process, in which the miners create more bitcoin by monetizing energy, is the most famous (infamous?) the other two components are equally vital to the operation of the system.

Consensus Protocols are the rules and guidelines followed by the nodes on a network. We have become accustomed to permissioned, centralized networks as a part of our Web2 experience. Centralized networks need decision-making parties, which means humans and committees—just the kind of thing that irritates the libertarian-leaning early supporters of bitcoin. It is that ideology of power and control by the few over the many that is challenged by the decentralized network and a big reason why they are so hard to create. Most people want more control, not less.

A perfectly decentralized network must have a pretty simple and strong set of rules. Because a network will need consistent behavior by its nodes to be able to execute anything useful, these wanted behaviors can be coded into the instructive code and updates shared by all nodes. There are *no managing humans* able to make adjustments or corrections; the network as a whole (or at least 51% of it) must agree on changes. This means that a decentralized network probably should not get too fancy with its code as complexity often leads to problems. A critique of Bitcoin is that it is slow and small in size, but arguments can be made that that is exactly what has led to its success. The few small bugs found in the early years were repaired long ago, and the 30,000 lines of code have seen only two major updates in the last 10 years. A relatively simple program has done very well in executing its goal: to be the first completely decentralized global network—permissionless, unsanctionable, and self-directed.

The consensus mechanism chosen by the founder of a blockchain is clearly very important to its longevity. It's not just about verification of transactions; it's also about control over network operations. If transaction approval is required by all nodes, the system will be comparatively slow. How do you decide who gets transaction approval, if not all nodes?

The set of rules for nodes that Satoshi assembled as the Nakamoto Consensus was written into the code as a permissionless way to verify bitcoin transactions and simultaneously produce new supply, known as *mining*. In connecting the verification process with the distribution schedule, an incentive was created: play by the rules, and you get a reward. These are the types of behavioral incentives that I believe have led to Bitcoin's success. This distribution schedule is predefined and finite, the result of which is the second intellectual innovation of an inelastic supply discussed in the next section.

Satoshi kept the code shorter by tying together a number of operations in the consensus process—verification, transaction retention, distribution of tokens and security. Many of the clever incentives built into Bitcoin's code rely on human psychology and asymmetries. When one cannot rely on altruistic human behavior, creating asymmetries can keep actors honest and provoke cooperative rather than destructive behavior. There are several asymmetries at work in the operation of Bitcoin. The cost to mine vs the cost to verify the mining, the rewards of participation vs the rewards of attack, the rewards from early participation vs the risk of waiting for more adoption, etc.

A·sym·met·ric Re·turn

noun

1. When the outcome of a trade has more profit than loss or risk taken to achieve the profit. Or, the upside potential is greater than the downside loss.

The Proof of Work protocol is just one of many "proofs" developed by computer programmers to satisfy distributed networks that their information is secure, that new transactions are properly approved, and that they are robust against attacks.

We are very familiar with online attacks. The TCP/IP rails set up for the internet are great for helping us to link computers around the world. They are not very good protocols for protecting our networks from attackers. Conversely, blockchain networks should be set up in ways that secure the data—Bitcoin has not been hacked in 14 years. Many of the other blockchains or cryptos have suffered outages and attacks. Why don't they just use Bitcoin's consensus protocol?

Two reasons. One, they want to *improve* on Bitcoin. They feel that the other elements of its protocols—the energy use, the permissionlessness, the blocksize, or the speed of transaction—can be 'fixed'.

Secondly, they want to have some kind of control over (and profit from) their creation. One of the most overlooked aspects of Bitcoin's Proof of Work model is that it is very democratic and very 'free-market' capitalist in its execution. It is not owned, controlled or manipulated by any person or group; it is globally decentralized. Other consensus mechanisms such as Proof of Stake are more replicative of our fiat system in that the more ownership (i.e., investment) you hold, the more you stand to make. In order to stake your ETH on the Ethereum blockchain, you must hold 32 ETH— roughly $65,000, or more than most Americans' entire retirement savings account. Other consensus mechanisms, in a very human way, usually allow for the inventors and early investors to profit by owning or controlling significant numbers of tokens or other command functions. I do not know that there currently is any other blockchain token or consensus protocol beyond Bitcoin's that can make the claim of full decentralization and governance by code alone.

The *Proof of Work* protocol, along with its operating partners the *Heaviest Chain Protocol* and the *Difficulty Adjustment,* work together as the Incentives Team for Bitcoin. Together they provide a defensive bulwark for the Bitcoin network that has proven successful for a decade and a half at discouraging serious threats.

Proof of Work requires a physical cost to be eligible to add new blocks of data to the blockchain. The Heaviest Chain protocol is a measurement that mining nodes perform to distinguish the chain version with the longest history—the most valid. This allows the miners to identify the "most agreed" version of the ledger, which is the one to which they want to add their proposed block. The Difficulty Adjustment keeps the system chronologically tuned the ten-minute block production goal.

The Proof of Work protocol has taken a lot of criticism for its use of energy. What is ignored is the critical link to real world costs that dissuades potential attackers. What makes this especially meaningful is that—unlike almost every other system on earth—the Bitcoin network has not been successfully hacked. This protocol has been operating autonomously for fourteen years, preserving over 800,000 blocks of data. *These don't seem like statistics that the business world should ignore.* Now that we are all using the internet, perhaps we should consider a platform that is proven safe for data.

We are so used to a world of controllers, issuers, managers, executives, that we find it difficult to understand why energy use is such a good enforcement mechanism. Of course, in the traditional world, the security apparatuses for governments and banks and corporations devour huge amounts of energy. I haven't seen any figures as to how much energy is expended by the entire financial system universe including counterparties, third parties, and governments to keep everyone all operating and integrated. The skylines of our cities are usually defined by the office towers of these financial giants. All these costs are justified in the name of 'keeping us safe' and our assets 'protected'.

Even those energy intensive pursuits that don't play a role in security are not questioned—video games and air conditioning come to mind.

In the U.S., it does not even occur to us that having 100% of our electricity needs covered 24/7 must result in some stranded or wasted energy—which was often not produced from renewable or 'green' sources to begin with. We aren't able to 'store' electricity in giant batteries somewhere—electricity created needs to be used or dumped. In the U.S., electricity is cheap and ubiquitous so we consumers waste it. Only relatively recently have manufacturers been identifying the most efficient energy product from those high-use appliances—and they are often ignored in favor of style or taste.

We fail to acknowledge that the legions of offices, staff, equipment, computers, and vehicles that we use in all our non-financial businesses require vast quantities of energy to operate.

At the same time that we are trying to convert our fleets of vehicles to electric, people criticize Bitcoin for being an electric network. As we save all our work to data centers, we don't want to admit that there is not a lot of difference in energy footprint or use case between Google's memory banks and Bitcoin's transaction storage mining facilities. We are now enamored with the idea of

Artificial Intelligence, which we think will be a huge boon to humankind...doesn't that run on electricity also?

Illustration 25: Very Good and Very Bad

We in 'developed nations' also choose to ignore other "bads" (as opposed to goods) created by our economies. We dispose of tremendous amounts of waste— everything we have produced but no longer have use for— which we pile into mountains at about 1,200 very large sites in the United States. We keep these dumps far from our residential areas, so they are out of sight (and smell) and mind. These facilities produce a tremendous amount of methane gas—a greenhouse gas more deleterious than carbon monoxide—which we also largely ignore. Only about 200 of these locations have any sort of remediation of this methane because the remaining sites are too remote to make either gas or electric transmission feasible economically. Bitcoin can help. Mining operations can be located on-site and do not need to transmit the energy anywhere. They can capture and burn the methane to power miners securing an electronic network. Economically, these miners can also provide local jobs and boost local demand for goods and services from the income inflows from mining operations.

The simple fact is that if you do not believe that Bitcoin is creating anything of value, there will never be a reason for you to think the network should use even one kilowatt of energy. This is an existential argument, not one of energy use. If your position is that you will never again need anything saved to a data center, then there is no reason for a data center to consume any amount of energy.

Increased energy use has been good for humans

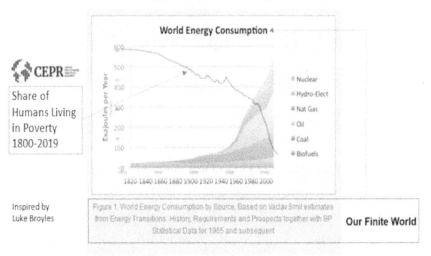

Illustration 26: Energy Use and Poverty. Charts overlaid to match the date range of 1800 to 2020.

Human development has always been dependent on energy usage. It is not the usage of energy itself that creates problems but the pollutants and wastes from the production and use of unsustainable fuels. There is a pretty clear relationship between improvements to human lifestyles and increased energy use, so we should be precise when we identify the problems with additional energy use. One would not know it from mainstream media coverage, but Bitcoin mining has now been shown to have the cleanest energy footprint of any major industry, with over 50% of fuel coming from renewable resources and economic incentives to continue increasing that figure.

TRUST THE SYSTEM, NOT THE NODE

A decentralized network of nodes that are programmed not to trust each other underpins the cooperative security of the Bitcoin network. One way that the effects of bad actors can be eliminated or reduced is by not trusting any actor in the network. To do so, the consensus protocol maintained by the network must be transparent enough for participants to be able to verify the validity of actors and transactions for themselves, without relying on the trust of any particular parties. The asymmetric construction of a commodity that is very difficult to create but very easy to verify emphasizes the 'social game' of maintaining honesty on the blockchain as a group effort.

Bitcoin's process combines this subtler incentive of total transparency of its workings with a specific requirement of the investment of capital—via energy costs—to such a degree that there is no economic incentive to attack. Instead of sanctions by 'authorities' to deter bad actors, these rogue attempts to infiltrate will be voluntarily identified and isolated by independent members of the mining community and left off the blockchain. Attacks will fail at the feet of thousands of networked miners, operating the same code to protect all users. Knowing in advance the cost it would take to strong-arm this system (it would take a nation-state effort at today's energy requirements) has prevented anyone from even trying.

The system as a whole balances itself and can be trusted to continue following its code, without the need for more specific client-fiduciary relationships used in the fiat world.

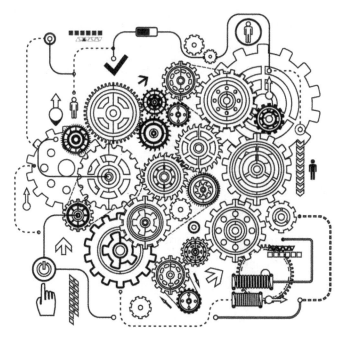

Illustration 27: Complexity

In the traditional world of qualified intermediary relationships, there are many overlapping systems that have been cobbled together and rely on trust between all parties. You may have a banker, an attorney, an IT department, an accountant, an auditor, and various payment partners, each with its own systems and security. The Bitcoin network integrates all these functions into one process, more simply and quickly.

Bitcoiners are also asked about the vulnerability of the network from within: what if "they" raise the limit of bitcoin? What if "they" change the consensus protocol as has done Ethereum, the second largest cryptocurrency? The threat of revisions to the code—or protocol changes— has also been minimized by design.

It is not easy to overemphasize the revolutionary nature of decentralization when we are so accustomed to *someone* being in charge. The libertarian roots of decentralization can also obscure its progressive goals. It is possible to celebrate both a small government ideology with a conservative protocol and progressive, democratic ideals of equality *at the same time.* To offer the freedom from control libertarians crave, with the equality and fairness sought by progressives. Bitcoin seeks to 'bank the unbanked', to offer an alternative to the permissioned traditional systems of value that exclude billions. Its creators realized that to survive, Bitcoin would need to have a slow but inexorable manner and approach to its tasks, not allow itself to be easily tinkered with and modified. This is best explained by Pierre Rochard of Riot Blockchain.

"Bitcoin's protocol changes require consensus from an undefined set of stakeholders. These stakeholders might include node operators, developers, miners, and users. Because there isn't a central authority dictating changes, achieving consensus among this vast, decentralized group is inherently challenging. This difficulty in making changes is actually a feature, not a bug, as it underscores Bitcoin's true decentralization.

Furthermore, the high threshold for consensus acts as a protective mechanism against hasty or potentially harmful changes. In any innovative ecosystem, there tend to be more ideas that sound good on paper but can have unintended consequences in practice. By demanding broad agreement before protocol adjustments, Bitcoin ensures that only well vetted and widely supported changes get implemented, preserving the network's security and integrity."

– Pierre Rochard

Humans have never before designed a system -analog or digital —that seems indefinitely sustainable in its operation and does not require any specific oversight. It is only now, in the digital age of certainty, that have we been able to use unemotional electronic tools to help us devise a benevolent and lasting network. We have Satoshi to thank for this entirely decentralized and simple, yet antifragile, design—an amazing invention for the connected world we have created.

PART II

A FINITE, PERFECTLY INELASTIC SUPPLY

FINITE SUPPLY

- What is a perfectly inelastic supply?
- Absence of an issuer and/or controlling party
- Money vs Currency
- Asset tokens vs Liability tokens: Elasticity
- Difficulty Adjustment and Halvenings—locking in the future distribution by time
- Automatic Disinflation, Proof of Work, Block Rewards —energy costs and rewards lock in the distribution by value
- Scaling with a Fixed Limit

This section is dedicated to the interesting concepts in Bitcoin that are associated with the idea of a fixed supply. Largely, this concept battles the idea that 'more is better', which is almost a natural law of the young (ask any kid if one cookie is enough!). When an organism or a species or technology is in its early growth phase, even evolution needs enough iterations to begin to produce mutations—kind of by definition. It wouldn't be a mutation if there was not an established baseline. Early development could be described as a "find your place and space" mode. Optimization is a process involving replication and repetition, hence volume. In its iterations and experimentations with alternatives, the growth phase always also involves a lot of destruction. Many things must fail to discover the one that works: this is a much more volatile environment. A growth phase will always end when the limits of optimization are reached, when specialization no longer brings efficiency improvements, and when that ecological or technological niche is filled.

At that point, the individual, or species, or technology must shift to a mode that could be better described as "maintain and upgrade" so that it stays relevant. The structure and bones may remain the same, but in maintenance mode, improvements can be made with the longer run in mind. Having found not only a niche, but also its limits, our organism needs to maximize its efficiency and longevity. This should be a much more stable phase of development that is intended to persist.

I see our fiat currencies as being in the "more is better" growth phase for humanity, and this concept of Bitcoin's fixed measure of value as being an indicator of a shift into humanity's maintenance phase.

WHAT IS A PERFECTLY INELASTIC SUPPLY?

Can you think of another 'good' or 'thing' that we have found or created that has a permanently fixed supply: that no more will ever be found or made?

No, you can't count things like the Mona Lisa or other works of art: the monetary premium for such a good is exactly in its originality. We'll call things like those collectibles—not without value, but not really used by 'people' in volume.

I'm talking about things that *many* people use. Can you think of a single thing we can't somehow make or find more?

I can't, and maybe that is one of the big reasons why the Bitcoin bug bit me when I read the *White Paper*.

A 'perfectly inelastic supply' is another way to describe, economically, the absolutely fixed amount of bitcoin that will ever exist. The fact that every bitcoin ever created has been, and will continue to be, on a predetermined production schedule that defines the total bitcoin supply forever is an entirely new phenomenon for our species.

> ## Elas·tic·i·ty *noun*
>
> 2. the responsiveness of a dependent economic variable to changes in influencing factors

Never before have we encountered an asset with an absolutely fixed supply. My favorite analogy here is with Manhattan real estate. You may have heard the phrase "they aren't making any more land" as a statement about its scarcity, but I maintain that it is an *illusory truism*. I challenge the reader to name any asset that has an absolutely fixed supply and remind them that Manhattan now rises hundreds of stories into the sky with created real estate. Technology allows us to do things in the future we don't know that we will be able to do now. The oceans are filled with microscopic gold particles and we just brought dirt from an asteroid back to earth. What natural asset is truly and definitively limited?

This iconic chart of the bitcoin supply schedule clearly defines, as executed by the code, how the supply of bitcoin will be mined into existence, regardless of the number of miners or price of the tokens. There are no variables to this schedule: it is fixed. It results in the creation of 21 million coins over a period of about 130 years, ending in 2140.

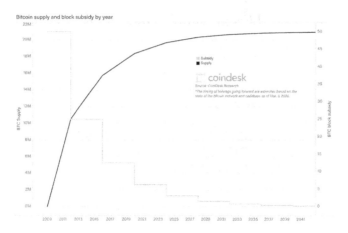

Illustration 29: bitcoin supply schedule

Because the Bitcoin supply is fixed, we can say it is *inelastic* in that a change in demand will not result in any change to supply. This would appear to create a highly asymmetric result in that the moderating effect on prices of more supply is not possible. We can't say this will happen for sure because adoption still needs to grow the demand, and we have not dealt with such a fixed supply in history— the future is always uncertain. However, as an academic economic matter, a growing demand and fixed supply should result in higher unit prices.

Several more things to note:

- The supply rose steeply in the first several halvenings (4-year cycles described below) and is flattening out. This purposeful disinflation helps to distribute a significant volume of coins relatively early, aiding in the adoption necessary to reach a Medium of Exchange utility. Although the rate of coins distributed halves every four years, the extreme divisibility of a digital asset allows for this halvening to continue another 25 times, or 100 years, or thereabouts into the future.

- The distribution schedule accounts for every smallest fraction of every bitcoin (called a Satoshi, or Sat) ever to be created, so while the rate changes over time, the distribution is guaranteed to occur as planned, leaving no "elasticity" in the supply curve; it is entirely predetermined no matter how people behave.

- The downward-stepping yellow line illustrates the declining amount of supply added over time. The disinflationary scheme dovetails with the Difficulty Adjustment, discussed below, and the overall supply limitations of the tokens.

Usually, variations in supply or demand impact a good's price upwards and downwards in a predictable, often self-correcting, way. Until now, there have been three variables - supply, demand, and price - whose interrelationships have helped us understand how basic economics work. Excess supply depresses prices, and shortages of supply tend to increase prices.

The classic example is that when the price of gold rises, miners have an incentive (and perhaps the capacity) to mine more. As more gold hits the market, and if the demand has remained constant, the oversupply will cause the price to fall again, and the miners will scale back the supply they produce. This equilibrium point between supply and demand always defines the price.

Recently, however, behavioral economics has shown us that we often make irrational or illogical choices and so the price itself may have an emotional, reflexive effect on suppliers or buyers in ways unanticipated by formulae.

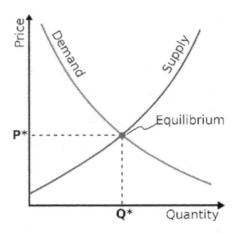

Illustration 30: Price Elasticity

This elasticity of supply—more or less of the asset is produced to balance demand—is not possible with bitcoin. Every bitcoin that has been and will be mined is on a predetermined course that will not change. This means that only two of the three pricing components can fluctuate—the demand for bitcoin, and, as a direct consequence, its price.

The early and basic economic lessons behave differently when the supply becomes a fixed quantity— even if it is still being produced and utilized. At any given point in time, there is a maximum supply that cannot be exceeded, which has changed our expectations of the future price of such an asset. If there is a maximum supply, then given an unchanging or increasing demand, there is a theoretical floor on price.

Here is a visualization of what an inelastic supply chart looks like, courtesy of Swan Bitcoin's research team.

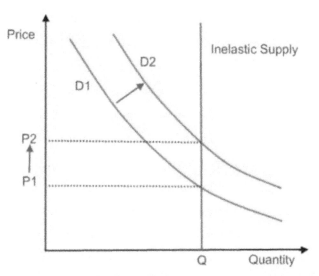

When demand D1 is in effect, the price will be P1. When D2 is in effect, the price moves up to P2. Supply is perfectly inelastic, so the quantity supplied to the market = Q for both demand curves. Any shifts in demand will affect only price.

Illustration 31: Price Inelasticity

Because there is no variance in supply, the usually curved line we saw in the last chart now appears as a vertical line.

Early bitcoin adopters take advantage of this inelasticity of supply by betting that adoption will increase demand and there is only one long term direction for price to go as no one will be 'discovering' any more sources of bitcoin. The adoption part of this bet can be questioned, but the math part is solidly a conclusion of traditional economic theory.

It is pretty clear right away that, on a planet of 8 billion people, a store of value (let alone a medium of exchange) needs to scale pretty well if it is to be adopted by any significant minority. Fortunately, these 21 million bitcoins are remarkably divisible (a million times more so than a dollar). We can create 2.1 quadrillion units at the smallest denomination (a Satoshi). There are enough units to handle global transactional business on-chain. As later mentioned, second layers to a blockchain can increase scalability, as do Venmo or ApplePay in the TradFi system. Bitcoin can scale *unit-wise*, an increasing coin price as valued by a free market will also allow for *scaling of value* as the stronger currency takes market share from weaker fiat currencies and other investments.

We have become accustomed to ever-rising prices under the controlled currency inflation scheme we have been running in the U.S. for over a hundred years. Prices rise because more money is created by the government and the banking system. The catalyst for money creation is that people or their government want to spend. Sometimes the people overspend and we have private sector recessions and recapitalizations. Sometimes the government overspends, which can have a much more existential impact. Often, as in wars or actions considered politically imperative, the money *has already been spent* via some sort of slippery arrangement to be 'rectified' in the future.

At the onset of World War I, a few people inside the British government concealed the fact that very few had stepped up to purchase the War Bonds being sold to raise money for the campaign. They ended up using public funds, lied to the press, and concocted a false premise upon which to build further support for a bloody war that killed 880,000 young Englishmen. Over 100 years later in 2017, the Financial Times issued an apology for their participation, unwitting as it may have been, for having "played its role in convincing the public that the sale was a success".

A loose supply schedule has usually been the end of a currency's dominance. An overspending government devalues its currency and may bankrupt the state. The more difficult it is for a government to increase the supply and debase its currency, the 'sounder' or 'harder' that currency is known to be. A commodity-money, or a currency strictly linked to a physically scarce asset is more difficult for a government to manipulate.

Figure 1. Major currencies priced in gold

Illustration 32: Money's Hardness

The following chart shows what has happened to the values of various modern 'government' currencies against gold, which has an annual issuance rate of about 1.5%. They lose value because they are inflating their supply faster. Gold is harder to produce than any of these fiat currencies. Note that even at gold's 'hardness', supply will double about every 35 years.

We usually experience inflation rates of higher than 2%, so this means that the value of your savings is halving even sooner than 35 years. When you are retirement planning, and have a forty-year career to fund a thirty-year retirement, realize that at even just 2% inflation, the value of the currency in which you are saving will halve twice during those seventy years. In other words, towards the end of your working and retirement years, the currency in which you are saving will only buy 25% of what it could buy when you started. Your target amount of dollars will need to 4x along the way.

In theory, over time, technology and progress should *drive down* prices. This does not happen in our system because we keep changing the measuring stick by adding more money. Prices rise in an ever-inflating fiat currency environment, but they don't have to. We have just become very accustomed to this over the last hundred years, four generations or so, and it is ingrained in how we think of money and prices.

Here are some price charts for shelter, food, and a consumer good to illustrate the point.

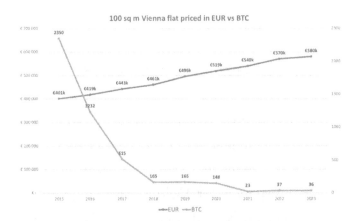

Illustration 33: The price of real estate in central Europe

Illustration 34a: A Big Mac priced in US Dollars

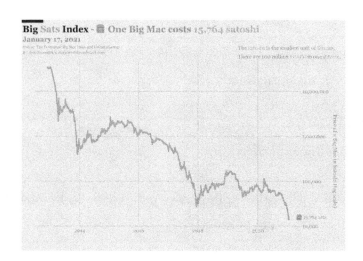

Illustration 34b: A Big Mac priced in bitcoin

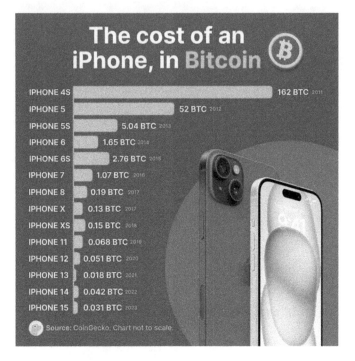

Illustration 35: An iPhone priced in bitcoin

In Dollars, the iPhone 4 models (2010) cost the consumer $199/$299 at the time.

The latest iPhone 14 models (2023) cost $799/$899/$999/$1099.

Is this price inflation or monetary inflation or both? ...hard to tell even for one product.

The prices should go down in bitcoin because the cost of production declines with technology and when measured in a fixed-supply currency, less is required to cover that cost. In other words, the difference between the increase in fiat price and the decrease in bitcoin price is the amount by which the dollar lost purchasing power to the bitcoin. The stronger currency wins—it is more desirable as a store of value. These charts probably reflect the debasement of the other currencies against bitcoin as well as the deflationary effects of technology, but the point is that this how prices for goods *should* look given technological progress.

It is strange to think of everything becoming less and less expensive over time instead of rising prices, but that is actually the true state of things. Things do actually cost less and less to make. It's just that the intermediate good (money) that we use to measure the value of these things becomes much more plentiful, and so there are more dollars in the money system for every Big Mac popped out of a McDonald's.

This is not a difficult idea; it's just tough to wrap one's head around because we have all grown up in a fiat currency world. It's a much easier thing to convince people that our money needs to keep growing because our populations and economies keep growing, but that is simply not true. Economies can grow even with a fixed currency—the currency becomes more valuable, so it needs to be divisible. Our systems grow the money supply by lending, but the other side of that trade, the borrowing, gets us into trouble so regularly that it is comical. That is the downside of using a growing currency.

We can recognize that things overall have become more affordable—for example, a middle-class family now might expect to fly together across the country to visit Grandma for Thanksgiving. That might have been out of their reach before commercial flights became inexpensive enough for most. When we think in terms of percentages of income spent on things, it can help us to understand the relativity of price and time.

Illustration 36: Global Free Markets

When money is created in our American dollar system, one party holds an asset and the other gets a liability. These entries can flow around our system as they are directed by peoples' decisions via the free market. Prices will rise in whichever sector is exciting investors at the time as these created dollars increase balance sheets. Large corporations and banks in the 80s, Technology firms in the '90s, Housing in the '00s, took turns as the darlings of investors as money flowed in and prices rose.

I'm not sure that we know how investment allocation and lending will evolve with a fixed money supply. I asked ChatGPT 3.5 for its assessment, here is its (his?her?their?) answer:

If the money supply is fixed, meaning that no new money can be created, the allocation of money among different sectors of the economy would depend on various factors, including demand, efficiency, and productivity. The distribution of funds would be determined by the interactions of market forces and the decisions made by economic agents, such as consumers, businesses, and investors.

Here are some key factors that would influence the allocation of money by sector in a fixed money supply scenario:

1. Market Demand: The sectors that experience higher demand for goods and services are likely to attract more spending and investment. If consumers, for example, prefer to spend on healthcare services, then the healthcare sector would receive a larger share of the available money.

2. Productivity and Efficiency: Sectors that demonstrate higher productivity and efficiency would likely be more attractive to investors and consumers. Productive sectors can deliver more output with the same amount of resources, making them desirable recipients of limited funds.

3. Investment Opportunities: The availability of profitable investment opportunities would influence the allocation of money. Sectors that offer better returns on investment may receive more funding than those with lower returns.

4. Government Spending: The government's allocation of its fixed budget can also influence the distribution of funds among sectors. Public spending on specific

sectors, such as infrastructure or education, can boost their funding and growth.

5. Consumer Preferences and Behavior: The spending choices of consumers can significantly impact the allocation of funds. If consumers collectively prefer certain goods or services, those sectors would receive more revenue.

6. Economic Conditions: The overall economic conditions, such as inflation, interest rates, and unemployment, can influence the relative attractiveness of sectors. Sectors that perform well in a particular economic environment may attract more investment.

It's essential to recognize that in a fixed money supply scenario, allocating funds to one sector implies diverting resources from other sectors. Therefore, decisions regarding resource allocation become crucial, and trade-offs between different sectors may arise.

Furthermore, with a fixed money supply, economic growth might be constrained unless productivity improvements offset the lack of new money. The overall size of the economy would be relatively stable, and changes in sectoral allocations could lead to shifts in employment, output, and prices.

I note that GPT answers are not perfect yet. In point 6, it mentions that "overall economic conditions, such as inflation," might influence sector allocation in a fixed money supply. In a fixed supply system, there would not be inflation.

The *growth constraint* fear mentioned by GPT in the last paragraph is the most frequently used defense I hear for the unlimited fiat supply we currently enjoy. Scalability through divisibility instead of replication is a difficult concept to grasp intuitively because we humans are still in the grow-or-die phase of our evolution. Now that we have established our dominance and mastery of this planet, we need to transition to our next phase of stewardship with an emphasis on management overgrowth.

ChatGPT is exactly correct in that it will be productivity improvements that offset the lack of new money. In fact, we already have product improvements; it's just that we are used to 'rewarding' innovation with higher prices to put money in the hands of successful entrepreneurs who create yet more money. If more money is left out of the equation, the best products command the best price, and everything else falls to zero in the long run. Please read Jeff Booth's The Price of Tomorrow (2020) for a beautiful description of how deflation might be a good thing. I agree that it seems strange that things—such as we know them today—don't cost anything. Consider that the letter you used to mail to your grandmother used to cost you at least the postage, if not the paper and envelope. Sent now as an email, the incremental cost to you is zero. We can assume you already have internet for other things, like the pen and paper for the letter, but the actual cost to you to transmit your writings is pretty much nil. Imagine that same transformation taking place with many, many activities in your life.

Consider a world far in the future where anything you want can be made and delivered by robots and AI. If the human population is balanced and our technology allows us to effortlessly produce anything we need, there is no need to work for money. Most of our basic necessities are simply created. What remains are art, collectibles, hobbies, and athletics: those things avidly pursued by amateurs for excellence. We could certainly occupy ourselves with whatever we wanted to take up our time; we just would not *need* to. How we decide to assign any remaining economic value to such products and services will be an interesting side story— perhaps that is where money will survive.

The last revolution of the growth phase brought us machines, which helped us conquer our physical environment and also made our lives substantially easier. Refer to the percent of humans living in poverty in the last two hundred years. The start of the mainte- nance phase for our species will involve nurturing our world through technology and artificial intelligence. It will help us conquer digital, information, and scientific worlds in the realm of the intangible. Blockchains, AI, and Quantum Physics, will also help make our lives much easier in the future.

As ChatGPT points out, there is really no distinct difference to investing with a fixed money supply; the key factors it mentions are all present already in the fiat world. Money is still lent and borrowed; it is just that the money lent already belongs to someone else—it is not created out of thin air.

When money's supply is fixed, the overall size of the economy may become more stable. This may be where the question of 'More v. Better?' is an interesting one. I think most will admit that there is a 'human carrying' population figure for the earth. At some point, we physically run out of space, and at a time long before that, we run out of patience with each other's proximity. So, if ultimately there is a relatively fixed number of humans living on earth, would growth cease to mean what it means to us now? For example, one can only really use one bed at a time. Sure, you can have multiple houses and multiple beds, but everyone has a point where another bed has zero value to them. Maximizing your return is then more about having a better, not another, bed.

Would we start to concentrate on 'better' over 'more'? I think so, and I think that is how a bitcoin economy would flourish in the distant future. When we are able to purge the idea that growth is everything, we will recognize that making things better is a more productive phase in human evolution. Just like many primitive insects have life cycles with quite different iterations of the same creature, we may as a species be approaching a cycle change when we morph into a more mature stage of being stewards and improvers of our world. This does mean that all things gradually get easier and cheaper as we reward the bringers of these qualities.

But because a fixed supply also means that ownership is a zero-sum game, it is easy to see that our old tribal prejudices and competitive zeal of humans will result in a distribution of wealth in some pattern that may resemble today's. Wealth will not be created, just redistributed with the fortunes of individuals. This may be where Bitcoin's strong social culture steps in and the globe decides how to allocate some of our wealth to the needy.

Consider the man who purchased 10,000 acres of what is now the San Fernando Valley. Was he a hoarder or a speculator or neither or both? I don't remember the details or from where the story came, but I find the metaphor appropriate. Of course, now this land has been divided and sold off many times into smaller parcels – just an acre today costs close to what the entire parcel cost a hundred and fifty years ago. The original purchase now consists of thousands of small homeowner parcels, with generations of families enjoying the land.

In a similar manner, the 21 million bitcoin will be distributed slowly by transactions and gifts through generations, as it simultaneously gains in value as an asset with a fixed supply.

History says that valuable, limited assets that can be subdivided *will* be subdivided.

ABSENCE OF AN ISSUER AND/OR CONTROLLING PARTY

Not having an issuer is a subject that can be addressed in all three sections of this book as it so affects each the decentralized network, the scarcity it creates, and the immutability of its records. In this section relating to inelasticity, or scarcity, being an issuer-less commodity means that there are no humans to 'manage' the code into changes that would affect the number of tokens created, the timing and contents of blocks, or the economic incentives underpinning the code.

Time blocking, difficulty adjustment, Merkle Tree pruning, cryptographic address creation, hashing algorithms, and other processes all work to allow the network to operate smoothly without any human oversight. The program follows its code through all these functions, taking care of all the duties that would fall under the Operations and Human Resources Departments of a tangible company. (Note that Bitcoin has no Sales, Marketing, Accounting, Legal or Executive Departments either!)

It is this relentless instruction-following that differentiates Bitcoin from all tangible businesses and nearly all online blockchain efforts. Its creator has found a way to manage a business via the crowd, instead of with any specific humans—a feat that has been confirmed by its fourteen-year track record at a 99.98% operational state. I call this a **digihuman system,** meaning that there is an underlying code that carries out the execution of the platform (decentralized and coordinated) but that humans (also decentralized and coordinated) retain responsibility for oversight, upgrades, maintenance, etc.

Illustration 37: Digihuman Systems

Carving up time into blocks helps the network avoid a classic problem of money: the double spend problem. Commodity-moneys largely did not have this problem, as they were identifiably and tangibly valuable. For a fiat or digital money, how do you stop someone from spending the same dollar twice? We have a long process to pursue those who bounce checks—and banks collected about $30 billion dollars in fees for it last year.

But that might not work with lightning-fast internet payments and settlements. Our fiat system conveniently allows for significant processing time for internal auditing and checks, and external synchronization with other institutions and the government. We have always lived with the delays of final settlement in our systems because there simply has been no other way to reconcile the number of accounts and transactions when the verification system relies on cumbersome balancing of conventional ledgers. This is not efficient or inexpensive, we just have not had any alternatives, until now.

The mining nodes are assigned 10-minute intervals to help organize transactions they verify as legitimate into discrete batches and work on solving the mathematic puzzle. Breaking time into blocks makes it easier to seek out illegitimate transactions in subsequent blocks of time and assists in thwarting classic network weaknesses. Although the time and transaction speeds of bitcoin are favorite complaints of many in the fiat payments world, we should clarify two items:

1. This system has resulted in virtually flawless execution for longer than any other network to my knowledge
2. Although they speak of "transaction speed", the traditional financial world does not want you to know about "final settlement", the time at which a transaction is deemed irreversible in their system, after bank audits, financial closes, balancing with federal reserve bank branches, etc.—which is about 60 to 90 days later.

When you send an "instant" Venmo payment through the existing financial system, that transaction is not officially completed for several months. Unsettled transactions are subject to cancelation for a number of reasons—disputes, lack of funds, audits, banking operations, etc.

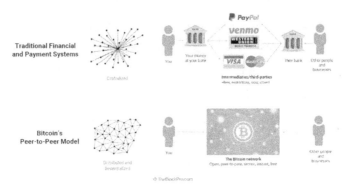

Illustration 38: Traditional vs Peer-to-peer

While it is true that most transactions have no problems, it is also true that they are not final for a lot longer than the authorities would prefer you to think. We cover over the highly inefficient processes in our system because it is easier to adjust rules and deal with exceptions than it is to effect a system-wide upgrade. Also, players that benefit from collecting fees for inefficiencies would never want to make income-threatening upgrades.

For example, there is no incentive for banks to lessen wire fees or Western Union to use the Bitcoin network because, despite faster and cheaper alternatives, their income partly depends on doing things the old way.

The Difficulty Adjustment mechanism operates about once every two weeks (at pre-set intervals of 2,016 blocks). This tool helps keep that 10-minute bitcoin block 'train' on time, which is important to keep token distribution also on its predetermined supply rate schedule. It does this in a relatively simple way, also integrated into the verification process and as a part of the Proof of Work consensus mechanism.

Like nearly all processes in the Bitcoin code, an independent node's operator is not overseeing his or her computer perform its work. The code automatically recalibrates the difficulty level of the math puzzle that the miners must solve to win a block reward by calculating the difference to the 10-minute block creation goal and adjusting to reacquire its target. The new calculation is also broadcast to the network to coordinate all nodes. This process happens automatically, as a part of the code followed by the node; an issuer or controlling party is not needed to make adjustments to the protocol's timing mechanism.

MONEY VS CURRENCY

In the prior chapter on the first decentralized network, we discussed the difference between physical currency created by a central authority and the idea of money, which includes credit and debt instruments denominated in a country's legal tender. We saw that, outside of real estate, most of the value created on earth is not associated with tangible, physical objects like gold and currency but instead with intangible creations like stocks and bonds (and, unfortunately, things like unfunded liabilities). We can understand that with the rise of financial markets, there has been an evolution over time of shifting value from the physical to the digital—a trend that seems likely to continue, another point touched on last week by Blackrock's Larry Fink. Even more recently, he characterized a bitcoin price rally as a "flight to quality", which is an entirely different compliment to the asset class. He should know, his firm manages more assets than any other.

Given that we are already comfortably using intangible value metrics like "full faith and credit" and their undeniable efficiencies, we should expect continued savings of time and cost, and any resistance to Bitcoin's digital-only nature should wane relatively quickly. We have already birthed a generation of 'digital natives', wedded to their electronic devices and navigating their ways through the virtual worlds they inhabit. While I remember a time before any personal computers, my adult children do not remember a time before smartphones connected the world.

It is logical, if not inevitable, that an absolutely constrained supply be created digitally. The physical universe is too vast and uncertain to guarantee something as being absolutely limited...or absolutely *anything* for that matter. It is through the minds of men and women that we can create and control a digital supply that is defined by its constraints.

This perfectly inelastic supply needs to live on the internet without a physical presence.

DIFFERENCES IN ELASTICITY

We can return to the concept of elasticity when we talk about different types of tokens. Generally, tokens which are assets will be somewhat inelastic—at least more so than your typical liability token. Asset tokens usually are, or represent, a bearer interest in the asset and so the only party involved has an interest in its value increasing over time. Liability tokens will always involve a third party, the debtor, who would like nothing more than the value of his debt (represented by the 'token') to decrease. A debtholder, such as a bondholder, must account for – and should price in - the many risks that might occur during the time of the loan.

For both of these types of tokens, an increased supply will result in a lowering of unit value; scarcity is nearly always valuable. However, the likelihood of changes to supply, another way of saying the elasticity, is much lower for a 'harder' asset like gold than a fiat currency. Gold's supply flow cannot easily be changed, whereas fiat currencies are easy to produce, especially when politicians are not restricted in their spending.

Asset tokens like gold coins or commodities are somewhat inelastic depending on their difficulty to produce. Fiat liability tokens are ultra-elastic.

Illustration 39: Fiat Monetary Inflation

Because the entire bitcoin supply schedule is predetermined, we can say that it is perfectly inelastic: it does not matter what actions are taken, the code has already determined exactly when each Satoshi will be mined (as long as there is at least one miner!), irrespective of economic conditions. Although the schedule is fixed in time, because the tokens are produced by the actions of an unknown number of nodes, *something* must act as a governor to synchronize block time with clock time. This is the role of the Difficulty Adjustment mechanism.

THE DIFFICULTY ADJUSTMENT

Although nearly all of the 21 million bitcoin to be created have already been mined, the process of continuing to mine is an important one because the creator has cleverly combined (for now) the operations of block creation and token distribution. In trying to plan for future growth, there must be flexibility in the system to account for unknowns, including the number of participants, and their timing. Satoshi introduced flexibility to Bitcoin by including a time-adjustment mechanism that works irrespective of participants or their contributions.

The token distribution schedule is integrated as a part of the transaction verification process. The block reward awarded to the miner who has won the right to attach his block to the chain is made in bitcoin. This is another process that the code executes automatically: it distributes the winning block to all nodes and sends newly-minted bitcoin to the miner. This is the only legitimate mechanism for the creation of bitcoin: the investment of resources into the system for a chance to be rewarded.

We need the miners to continue processing transactions to create more bitcoin. At the same time, we need this distribution mechanism to ensure that the *rate* at which bitcoin are created stays within the parameters of the overall plan.

This is where the Difficulty Adjustment plays a its critical role —essentially, it locks in future distribution by carving up time into ten-minute blocks. As long as the code can keep the distribution 'on the clock', the token quantity produced will be guaranteed to match the quantity planned on the schedule.

This is brilliantly done in a simple way: by varying the difficulty of the mathematic puzzle the miners must solve. If the puzzle is too easy for the number of miners at work, they will resolve blocks quicker than the goal, and the puzzle difficulty needs an adjustment upwards. Conversely, if the miners are taking too long, the puzzle difficulty is adjusted downwards. This is not a terribly difficult puzzle; it is more like a kids' guessing game. The code is thinking of a number between 1 and some number – what is it? The lower that number the fewer possible correct answers.

This requires the ability to just process as many guesses as possible – it's not the same type of intelligence as AI, not smarter thinking, just more guesses being made.

This process has been happening every ten minutes since 2009. After each 2,016 blocks are mined, roughly every two weeks, the Bitcoin network checks its clock. If the network is 'on time', no adjustments are necessary or made. If there has been a change to the hash rate—the computing power brought to bear to solve the puzzle—and the timing is off, the network will modify the target range to make it harder or easier for the miners, which slows down or speeds up the network's processing speed. This flexibility allows for real-life variations to occur, such as changes to the numbers or power of the miners, without disrupting the network's overall timing and distribution goals. This Difficulty Adjustment is another innovation buried in the code of an autonomous network that makes it possible to execute a plan over a very long period of time, without full knowledge of participants or circumstances, and allows for a perfect inelasticity of supply. The Difficulty Adjustment "keeps the train on time", locking in future token distribution by time.

THE HALVENINGS

The inventor of Bitcoin could have chosen a linear formula to generate new coins via block rewards, which would have created a consistent number of new bitcoin every ten minutes for the entire distribution period. This would have worked from a logistical standpoint, but Satoshi realized he was dealing with humans and our emotions and that he needed to account for our behavioral tendencies, too. In order to become a Medium of Exchange, a money needs to be in enough supply such that sufficient numbers of users can create enough momentum to 'prime the pump' to scale up usage into meaningful quantities.

Satoshi created an asymmetrical pattern that started out distributing 50 coins per block, but which rate declines every 210,000 blocks, or about 4 years, so slowly over time. This allows for bitcoin to be mined in continuous, albeit diminishing, amounts for the next 120 years until it reaches its predetermined limit of 21 million. (What happens in the year 2140 when the last bitcoin is mined?

No idea. But miners already collect fees as well as the block rewards, so likely a higher fee structure will be phased in at market rates.)

$$\frac{\sum_{i=0}^{32} 210000 \times \left[\frac{50 \times 10^8}{2^i}\right]}{10^8} \approx 21,000,000$$

Illustration 40: The Halvening

The Halvenings[1] component allowed the network to supply the market with sufficient quantities at its outset to give the technology its best chance to survive. Satoshi imagined that any network effects needed an early boost, after which the distribution could wane.

Along with assisting to jump start bitcoin usage, the halvenings represent a restriction on supply on a regular and pre-known schedule. This has led to interesting price cycles as it has an effect on the number of bitcoin that can be brought to market at any one time. Together, the Difficult Adjustment and Halvenings help allow the bitcoin network to function autonomously, as pre-determined guardrails, to allow for an inelastic supply to be metered out over time.

AUTOMATIC SLOW DISINFLATION

From here on, for about another 120 years, this regular halvening process means there is a built-in slow disinflation—a declining rate of inflation over time—as illustrated in the supply chart. Satoshi has designed an elegant way to both meter the ongoing supply and provide an economic incentive to participate in the network. For the purposes of our lifetimes, bitcoin supply will be ever expanding at a slow rate to end up with a finite and fixed final supply of 21,000,000.

This very low but ongoing inflation rate may help to satisfy investors and debtors alike. Value is kept removed from the most insidious of all threats to currency: monetary inflation. Although it will be generations before the asset supply becomes permanently fixed, the next halvening will drop the new supply to half that of gold, the only other asset globally employed as a currency (never mind the inflation rates of any fiat currencies!) At the same time, modest income may be made by lending to the benefit of debtors in need of funding for their own purposes.

The halvening also has an effect on mining economics—every four years the ranks of miners are severely tested when their revenues, as represented by bitcoin block rewards, are halved. Depending on the amount of energy invested in mining at the time, or hashrate, and the market price of bitcoin, the miners' profitability is severely tested and many miners are forced to exit the business. This turnover in low-profitability miners every four years is another example of how subparts of the innovations are themselves revolutionary in creating economic and behavioral incentives that not only grow the network, but do so in an efficiency-minded manner that rewards work, investment, and patience. The flushing out of poor performers is not an action performed by the code or users, but is an economic consequence of a truly free market operating with an immutable set of rules.

PROOF OF WORK: VALUE VIA ENERGY

Much as the Difficulty Adjustment effectively 'locks in' the future distribution schedule 'by time', the Proof of Work consensus protocol serves to lock in the future distribution 'by value'.

Although it is vital to the security of the blockchain as Satoshi noted, Proof of Work also provides a real world economic link along with a scarcity which creates value—or so say the buyers of bitcoin. There is a link between a commodity's cost to produce and its market price. As mining becomes more competitive, its costs go up.

Proof-of-work has the nice property that it can be relayed through untrusted middlemen. We don't have to worry about a chain of custody of communication. It doesn't matter who tells you a longest chain, the proof-of-work speaks for itself.

– Satoshi Nakamoto

In order to make an impact on bitcoin's verification process (in either a participatory or malevolent way) one must invest in the acquisition of energy. The energy required to compete in the block-approving mathematic problem is significant, representing a real-world investment on the part of all miners, not just the winning ones. The higher the total computing power employed simultaneously by all nodes—or hashrate—the more difficult and costly it becomes to be a bad actor trying to influence even one block. One can see that increasing the security of the network, which is what this is doing, should transfer some economic value to the underlying tokens.

Miners are incentivized economically because the math works (most of the time). They can sell any bitcoin they receive as block rewards for more than it cost them in electricity to mine. However, at times during cyclical bear markets in bitcoin's price past, a majority of the miners have been unable to survive the price declines. It is a very competitive and capitalist (American!) market.

You may also be able to see that if you can get access to cheap electricity, you can make good money (and maybe even survive the next bear market). That is why you hear about bitcoin mining with alternatives like wind, water and solar—and even volcano energy. It's a real winwin when a bitcoin miner buys power that would be "thrown away" by the grid to balance its loads. Or when a miner can connect to a landfill's methane gas and turn a pollutant into a productive energy source. That's cheap and available energy that bitcoin miners are seeking out for the benefit of all of us.

And despite all the FUD (Fear, uncertainty, doubt) about its energy use, according to the Cambridge study, bitcoin mining still does not equal ½ of 1% of our demand and is now the 'cleanest' industry in the global economy. Ironically, although Bitcoins purpose was never environmental in nature, it is one of the cleaner technologies we utilize, making these criticisms not only inaccurate but so unaware and hypocritical they seem childish.

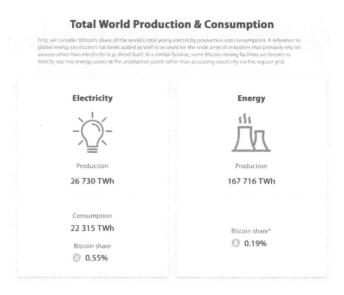

Total World Production & Consumption

First, we consider Bitcoin's share of the world's total yearly electricity production and consumption. A reference to global energy production has been added as well to account for the wide array of industries that primarily rely on sources other than electricity (e.g. diesel fuel). In a similar fashion, some Bitcoin mining facilities are known to directly tap into energy assets at the production point rather than procuring electricity via the regular grid.

Electricity

Production

26 730 TWh

Consumption

22 315 TWh

Bitcoin share

0.55%

Energy

Production

167 716 TWh

Bitcoin share*

0.19%

Illustration 41: Energy Use

Bitcoin's overall scarcity has continued to work to keep its market price above its mining costs. A social culture of holding onto purchased coins (see HODLing or Stacking Sats) along with issues like lost access, political restrictions, and macroeconomics have contributed to an overall rising price chart for most of its history (best performing asset YTD 2023).

This success, especially during more manic periods of its cycle, has led to continued aggressive price targets. Which begets more innovation and growth in all the sciences ancillary to the participation in the world's first asset with a perfectly inelastic supply.

SCALING WITH A FIXED LIMIT

Many people feel that only 21,000,000 coins will not be enough for the human population who wants to use bitcoin. It can be hard to fathom just how divisible a bitcoin is. A US dollar can be divided down to pennies, or 1 of one hundred.

A single bitcoin can be divided all the way down to a Satoshi – one hundred millionth (1/100,000,000) of a full coin – much smaller than a US penny. Right now, a single US dollar can be exchanged for 3,780 Satoshis – you don't have to own full bitcoins. The ability to economically transact micro-amounts is only possible in the digital world and may become very useful. So if there are a total of 21 million BTC, there are 21,000,000,000,000,000 Satoshis. That is enough for every single one of 8 billion humans to hold 262,000 units, so there are enough of them for a long time. And if the time arrives when there needs to be more divisibility (not more bitcoin) what better group than a dedicated band of program-mers who must cooperate on shared goals to accomplish the task?

The other thing easily overlooked in a fiat world is that that value of each coin can also rise in tandem with the amount of wealth for which it is traded. Because of blockchain transparency, we can see exactly what price was last paid for each bitcoin, and the aggregate price paid for all tokens may be a better measure of value stored in the technology than the various pricing metrics used in the equity markets. HODLing bitcoin as a store of value against a fiat currency that has declined in unit value for over 100 years straight, is attempting to leverage the scarcity of an asset with an entirely inelastic supply.

PART III

AN IMMUTABLE AND VISIBLE LEDGER

IMMUTABLE LEDGER

- Cooperative security
- Energy secures the ledger's immutability
- Contracts, Performance, Auditing
- Permissionlessness in a permissioned world

When there is sufficient transparency and decentralized verification processing, it becomes possible to create and maintain a ledger of account that is free from manipulation and theoretically permanent in nature. In order to be able to say that this is the first time we have created such a tool, we needed to clarify how we define "permanent" and immutable. I say this because we do have ancient stone carvings that still exist today, far outlasting their civilizations and so permanent in a way. We can't be sure that someone did not corrupt the chiseled inscriptions, but carving them into rock seems pretty immutable too.

When we enter the digital realm, though, we know immediately how easy it is for people to create replicas and corrupt files or 'hack' networks for all sorts of purposes. In the quarter century that the general public has had access to the internet, we have come to understand that it is not a safe place. Being permissionless has been a big reason behind the internet's adoption curve. It also obviously exposes all users to all users. Security was not the goal behind the creation of the internet, connectivity was. Now we need tools on the internet rails that can also provide us a measure of security.

Satoshi's work on a peer-to-peer currency solved many of the security challenges inherent in the TCP/IP setup of the internet. He or she chose this cooperative, "many eyes", format for general network security, which has not been breached in its fourteen-year existence. I cannot think of anything else in the digital realm that can make such a claim.

An immutable and transparent ledger represents both a contemporary agreement by the users that its transactions represent the consensus of "the truth," *and* it demonstrates its ongoing trustworthiness by allowing total access to its pages by anyone, forever. I can't think of another example of an entity that keeps every detail of its transactions permanently available for the public to view at any time.

Illustration 43: The Bitcoin Train

It can be difficult to envision an intangible process. The image that helps me most is that of a train, slowly but inexorably chugging forward, attaching one new car every ten minutes. The cars are full of the time-apportioned transactions that have been verified by the miners and are linked together for security. It's easy to tell if any car has a problem: it affects every car behind it.

This section is dedicated to the interesting concepts associated with an immutable ledger, the third and final transformative element in the trio of innovations. Ledgers have been in existence for millennia, so arguably a ledger should not be innovative. My stance is that a *distributed, permissionless* and *transparent* ledger is actually a whole new thing, worthy of being called transformational. The reason I see it this way is because it helps redefine who controls history. The old saying about history being written by the winners is very true, but Bitcoin redefines that framing. Now, history can be a "truth" agreed-to by everyone, voted on by the crowd and preserved in perpetuity.

To me, this is a sea change in the traditional power and control humans have wielded over each other. With these new technological tools, we can democratize governance, introducing such new practices as truly distributed and cooperative security.

COOPERATIVE SECURITY

There are many examples of cooperative security in the natural world. Schools, herds, and swarms are very effective in their biological niches as defensive mechanisms from interspecies competition. We as humans have made mutual defense our practice, in a tribal way, probably from our beginnings.

Having 'many eyes' on an asset has proven effective for humans in the physical world, but also as bitcoin's security net: to disallow by plurality vote any action that would violate the operating code.

An interesting side note to this is that if a significant minority really wants to 'break away' from the primary user group, they can. They are free under permissionless protocols to modify their code in a synchronized way and create what is known as a fork in the blockchain. One chain becomes two at a certain blockheight, and the two versions of the code proceed to create their own blocks of information on respective chains while sharing a pre-fork history.

In the natural world, this might be the equivalent of a herd splitting into two groups based on the members' assessments of their existential prospects in following one of two self-appointed leaders. Cross the river to the lowlands or remain in the higher elevations? There are no guarantees that the majority will make the best choice.

Time and experience will determine which resulted in the optimum outcome. As long as participants are free to decide, the strongest code, that which the majority perceives as being the most useful, will likely survive.

The forking aspect of blockchains seems to me to be very democratic in that—given permissionless participation—the most popular version of that protocol ends up winning, and most forks end up having their tokens 'crash' in value against their original as the new ideas rolled out in the fork just didn't take, or were a disingenuous copy by a fraudster.

Bitcoin has been forked a number of times into new blockchains: Bitcoin Cash, Bitcoin Satoshi Version, Litecoin, etc etc. None of the forks have gained anywhere close to the value or the market cap of the original. They have small communities of users who seem to be content with their versions.

A fork can also be intentional on the part of the community when most nodes agree to modify their code to a new standard and make the same changes to their code. This can be done to effect upgrades, patch bugs, prevent attacks, or other defensive needs, allowing the network to remain 'alive' in its ability to adapt to change. Essentially, the community creates a change in the code, and all nodes voluntarily make the change within a time window to send that blockchain down a new "track" in terms of its operation. There is no second track created and no competing tokens; the original track is just modified. It has to be a pretty obvious, basic and popular upgrade to be approved separately by tens of thousands of nodes, so in a truly distributed network like Bitcoin, forks are rare.

Illustration 44: A Blockchain Fork

Bitcoin has been intentionally forked several times to allow for upgrades that have increased useability, like allowing for multiple signatures, additional information in header fields and other improvements that have been approved—with great effort—by the community. It is likely that there will need to be a security fork when quantum computers threaten Bitcoin's cryptographic processes.

In the short term, the code and its incentives play the role of security for the blockchain platform. People and software keep the verification process and the transactional, active part of the platform operational.

In the long term, *energy* secures the ledger's immutability.

ENERGY SECURES THE LEDGER'S IMMUTABILITY

Energy use is now the most used objection to bitcoin, whether this is from simple ignorance or malevolent FUD (fear, uncertainty, doubt). We have previously mentioned how a real-world cost dissuades so-called 51% attacks on the network. No other network to my knowledge has survived the last decade without being hacked or penetrated in some way, which should really be the headline for the story about energy.

Complaining about bitcoins energy use is also <u>choosing to ignore a number of material facts:</u>

- In 2022, **$8.2 Trillion dollars' worth of transactions** were done in bitcoin, and the network currently stores nearly a trillion dollars of wealth. Supporters would argue this is a productive use of energy, especially as compared to the societal benefits of other uses such as:

- Video Games
- Streaming TV
- Google storage space
- Twitter, Facebook, Instagram etc, who are arguably providing value for their shareholders
- The amount of electrical use by 'scientific endeavors', many of which are poorly done and will be failures. No one says "stop the experiment on fusion energy because it wastes so much energy". Even malinvestments in R&D are considered to be worthwhile efforts to find better ways.

Illustration #44

Illustration 45: How We Live

- How we live in today's world does not reflect an overall reduction of energy at all. We do not seek to save energy for any reason. No one is going to walk from LA to NYC to save the energy and pollution from the airplane. Automobiles, refrigerators, plastics, air conditioning, physical waste are all areas in which we could make major changes if we cared, yet we do not.

- Playing during the daytime instead of lighting up major sports stadiums at night is a trade no one has offered to make.
- We are so conditioned by our surroundings that many of the common examples of wastes of electricity do not occur to us; they are just part of the system in which we live, so we do not notice.

Unless small groups of critics use hyperbole and inaccurate statements to stir emotion and create an outsized and inappropriate negative impression, we don't notice a lot of stuff:

Standby Power: Devices and appliances that remain plugged in but are not in use still consume electricity in standby mode

Inefficient Lighting: Traditional incandescent bulbs waste a significant amount of energy.

Over-Illumination: Using excessive lighting in indoor and outdoor

Leaving Lights On

Inefficient Appliances

Overusing Air Conditioning/Heating

Phantom Loads: Devices draw power even when turned off, such as devices with remote controls or clocks

Inefficient Industrial Processes

Over-Refrigeration

Charging Devices Unnecessarily

Excessive Water Heating

- Many people are not aware that we overproduce for the amount of gas we use, flaring (literally burning off) the excess gas at a rate of a little over 1.25% of total gas production in 2018, 342 billion cubic feet (Bcf).

Some 40% of our electricity comes from burning gas. It doesn't matter where this electricity is used—to charge your car, make your toast, or mine your bitcoin, it comes from the gas and so attaches to this flaring environmental cost as well as any other environmental damage caused by the original mining of the gas.

- Another example of how we currently live (without questioning how) and how we can improve that with bitcoin mining has to do with landfills. Bitcoin mining can create economic value out of pollutants by burning off the methane that seeps out of the 1200 very large landfills across the country. It is not feasible to send the gas to urban demand areas, nor to transmit the electricity from an on-site generator. It is, however, economically efficient and environmentally beneficial to use the methane to power bitcoin miners on site.

Interestingly, in this case, the bridge bitcoin creates between the physical world and digital economic value *works in reverse*. For the last fourteen years, every moment, the investment of a real-world asset—energy—into the Bitcoin network has grown the network's value by securitizing and protecting its integrity. With landfill mines, the miners return value to the real world landfills by creating digital assets to reward the real-world work that reduces their heavy pollution. In effect, the landfills create a public liability, and the bitcoin assets that are created in mitigating the liability during mining serve to offset that liability.

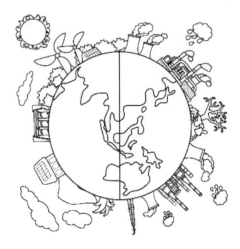

Illustration 46: Energy, the global asset

There are always groups who will be opposed to new technologies. In this case, a clear and lasting legacy of this technology is the ability to secure and protect value in ways that have not been done before. Securing value with energy is not a new idea.

Henry Ford had a vision that energy could be used as the new basis of a new economic system – a currency. He believed that linking the currency to real world productivity could help stabilize economies and prevent financial crises.

In his book "Wealth, Virtual Wealth and Debt," published in 1926, Frederick Soddy argued that the prevailing monetary system based on gold was flawed and proposed a new monetary system based on the concept of energy. He believed that wealth should be measured in terms of energy and that the value of currency should be linked to the available energy resources in a country.

In his book "Critical Path," published in 1981, Buckminster Fuller also proposed the concept of an "Energy Currency" as a solution to global economic challenges. He too argued that the true wealth of a nation was its energy resources.

Along with contributions in the last thirty years by Adam Back, David Chaum, Wei Dai, Hal Finney and others, the release of Bitcoin in 2009 marked the actualization of an idea with a hundred-year history. Bitcoin is not a new concept, it's just taken us a while to come up with all the pieces in one place. Thanks Satoshi.

CONTRACTS, PERFORMANCE & AUDITING

Most human interactions boil down to agreements. How we agree to act, behave, live, dress, earn a living, transact with others, and so on. Most of our business world is made up of contracts to represent underlying agreements to do or make goods or services. There still needs to be a lot of trust. You need to trust that the other party is who they represent, and they need to perform per the contract. Of course, not every agreement is properly fulfilled, so there are a number of systems we put into place to help. We have verifications and credentials at the front end, and we have the legal system as a form of redress should performance fail to meet the agreed standard.

As we have seen, blockchain technology really excels at providing verification via cooperation. It can play a role in actual value transfer as a store of value or medium of exchange. It can assign scarcity to intangible tokens or link digital coins to real-world assets. It reaches a final settlement in less than an hour without intermediaries. It preserves its transaction history perfectly through time, accessible to anyone, without the need for permissions. In short, it really excels at executing the logistical portion of a contract between parties.

We are exploring how a blockchain can be modified to carry information other than just simple input-output numbers. A smart contract is simply a small computer program representing its parties' agreement, saved onto a blockchain-style database. Because the contract can have self-executing bits of code and blockchains track time closely, the contract may actually remain dormant until some specific time or event 'wakes it up'. As you read this, developers are working on more and more smart contract capabilities. It seems safe to say that one day, most multiparty contracts will not live on pieces of paper but will be smart contracts on a blockchain.

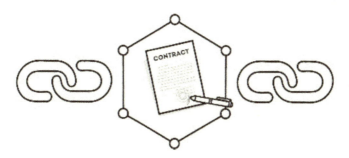

Illustration 47: On-chain Smart Contracts

In terms of disruptive ideas, the transformation of contracts over the next decade will be monumental. Our economy employs many people to track items, verify people, audit results, execute a transaction on a particular signal, and sanction those who don't conform to standard. All these agreements to do things can be transferred to the digital world and auto-executed. Smart contracts will revamp the business world. Bitcoin has no auditors.

PERMISSIONLESSNESS IN A PERMISSIONED WORLD

As mentioned, our current systems have layers upon layers of protocols, each requiring separate permissions. Just to operate this computer, I need to remember my Operating System Password, Screen Saver Password, Google login password, and email login. I subscribe to a password app for security. It creates and remembers passwords for me. I've only had it a few years, but it manages several *hundred* passwords for me.

Being permissionless in a permissioned world is a pretty big deal. Everyone has had an experience of being excluded or asked to leave, whether it's the social media thread, the local library, the neighbor's pool, the big kids' group, the members' club, the restricted area, or the moms' newborn group (no dads!) Having a place that does not require some kind of authentication, verification, membership, permission, or entry fee is an increasing rarity. To some, it feels like an essential freedom.

Quite clearly, security (the internet's Achilles heel) is a big deal as always, and Bitcoin seems to have it figured out—we shall see about the altcoin world.

Another difficult-to-resolve challenge may be how the crowd finds a balance between free speech and a level of civility of which most would be in favor. Is there such a thing as polite censorship by the mob?

IDENTITY AND OWNERSHIP

An immutable ledger can also finally allow us to preserve and easily access our identities and ownership records without the need for constant analog, intersystem, manual reverifications every time a new service or procedure is requested. Quick Read (QR) codes can carry a link to various iterations of our identity for instant verification of nationality, membership, health care, ownership and other records on which we currently spend unknown amounts of time on applications, processing, and approvals—from the line at customs to the DMV to the bank. We should not need to be constantly identifying ourselves with outdated paper tokens. It should be easy for someone to verify ownership of a material asset.

Identity theft has caused us untold millions in losses. We now have ways to preserve privacy and equip citizens with identification protocols that allow us time for those hobbies and activities we really enjoy.

CONCLUSION

AN INTERESTING POINT

At the time of this writing, we are at a very interesting point in Bitcoin's history. By the end of July 2023, we have seen a lack of any clear guidance from most world governments on any aspect of the digital world seemingly suddenly unlocked by bitcoin. We have gone through successive cycles of boom and bust in price, while a relatively steady adoption in mining and usage has outlasted all challenges to its permissionlessness.

Yesterday was the 15th anniversary of the release of the White Paper online. While that seems like a short time for a new value system to have been alive, it is a very long time for a new value system not to have died!

"It might make sense just to get some in case it catches on."

– Satoshi Nakamoto

Though relatively very small, bitcoin has been the best-performing asset for the last decade. This has gotten the attention of the financial and investment worlds more than any other factor. Wall Street certainly does not like to see anyone make more money that they do. The institutional involvement and investment that we anticipate in this next Bitcoin cycle after the 2024 halvening may end up driving retail adoption and pubic sector regulation more than any other factor. The institutions are coming.

REGULATORY CHALLENGES

There is no question that governments have not been quick to understand, let alone regulate, the world of digital assets. Their very variability and flexibility make it difficult for regulators to categorize neatly as in the world of orange groves and Howey Tests. There seems to be a reluctance to admit or understand what opportunities these new tools present. Instead, there is a focus on sanctioning behavior that does not conform to rules written for more static real-world assets with long-established histories of oversight.

Having said that, at the moment, It is not clear to me that any other inventions in the crypto world fit ANY ONE of the three innovations brought us by Bitcoin. Regulators have a point when they criticize the crypto space as a Wild West of money and ideas. This might not be so, however, if rules had been made early on that corralled (if you will) the overexuberant hucksters and made clear to all how to operate legally.

This is all a part of every new technology ever, though, so it may not be curable. Whether it is automobiles or telephones or internet companies, there is always fraud, and there are always those trying to take advantage of people's weaknesses to sell things that are just not so. We have seen a great deal of value vanish with the exposure of the frauds in 2022 and the concurrent bear market. Perhaps a more rigorous but fair approach on the part of government agencies might have protected more of the investors they claim to be defending.

The biggest technical question Bitcoin continues to answer is most succinctly voiced by Parker Lewis: Can Bitcoin continue to successfully defend its fixed supply?

The biggest question facing the inevitability and timing of a future with bitcoin (and how well altcoins are controlled) may be the degree to which the United States' legislators and courts are going to participate in these changes.

There seem very legitimate reasons to consider some of the offerings in the crypto world as being securities, even under a layperson's perspective, like mine. The recent ruling in the Ripple (XRP) case may have tried to distinguish between early venture capitalist participation in a project (owning a share of an orange grove) and secondary market purchases of the created tokens (buying an orange), but these concepts need better clarification and guidelines.

SEC Chair Gensler seems more concerned with political posturing and agency power than in actively exploring and encouraging the development of a new asset class. He could leave a legacy of being the architect of sweeping updates to financial codes, finally admitting the digital era through the gates of the entrenched financial system, fifty years after the introduction of personal computers.

But especially for someone with prior knowledge of the potential of blockchains, his behavior seems petty, controlling and myopic—certainly not what anyone would call inspirational in the context of transformative technologies.

Our regulators need to work through the differences between raising money and control of protocol, or utility tokens and store-of-value coins, or public and private blockchains.

Illustration 48: Regulators

As easy as it is to see that many of the projects in the 'crypto' world appear to be securities on their face, it also seems a no-brainer that the digital and programmable assets we are creating now have properties and capabilities that do not exist in the tangible world. All of the financial rules in place are based on tangible items, and even if their derivatives offer exotic flavors of risk or calculation, the underlying real-world asset and its capabilities are the starting places for any value measurement. Digital assets offer us characteristics, because of their programmability, that we simply cannot offer in the real world.

Traditional physics says that a bit of information can be represented by a 1 or a 0. Quantum physics, the newest interpretation of our universe, proposes that a 'qubit' of information can be represented by a 1 and a 0, at the same time. This seems as non-sensical as a magical internet money, but scientific research is relentlessly pushing us to resist our intuitions evolved for the macro physical world and accept that things might not always work the way they seem. Remember when the earth was flat? It still seems pretty flat; some people still believe it is.

SCALING

This phenomenon of needing to revise our understanding of even hugely material subjects has happened again and again in history. When you shift into a new paradigm, the old rules are often useless and survivors are the ones who adapted. Much like the exploration of quantum physics, our scientists are currently working through exactly how decentralized networks function, and how matter, energy and information may be interchangeable.

Today we don't suffer from slow scientific advancement, lack of funding or technology difficulties, we suffer from a lack of direction because our capabilities are expanding much quicker than our leaders' abilities to understand them.

Our political leaders can influence greatly the speed at which adoption of a useful technology occurs. Historically, they cannot stop the spread of something people generally consider to be useful, but they can slow its distribution in a variety of ways.

Our business leaders can also affect the adoption curve by virtue of their own personal conversion, leveraged by the size of their organization and the business network to which it attaches.

Swan Private Insight, Issue 22, April 2023:

"Many people in traditional finance, academia, the media, and the government believe strongly in these two flawed narratives:

1)"Managing the currency and monetary system must be done by the state."

2)"Creating new units of money is necessary to achieve economic growth."

Today's media reaches millions, for better or worse. We can see how groupthink is a very easy habit to fall into when algorithms channel human emotions into narrow streams. Changing people's opinions about something can take a long time, even something clearly useful. When opponents mischaracterize and frauds sow confusion, people's resistance to change only stiffens.

The snippet, courtesy of Swan Bitcoin's community (author unknown) seems to me particularly astute in identifying underlying beliefs that are hard to shake.

In some senses, teaching minds to be open about new knowledge is the biggest challenge. This is ironic given the standards to which we claim to hold education and wisdom.

You may hear that Bitcoin the network is slow and too small to scale to the uses bitcoiners project. By now I hope you realize this (aside from being misleading) is intentional in the part of its inventor. He or She was certainly wise enough to see forward—he made remarks about the durability of his ideas over time. Knowing that it can take a long time from introduction for a disruptive technology to spread, the inventor employed game theory to reason out how users and attackers might behave. In his/her writings, we know that Satoshi was intent on future-proofing his ideas as much as possible to give them the longevity he/she felt they needed.

There are two main scaling complaints: 1) That there are not going to be enough units in the world for 8 billion people and 2) that it is too slow.

The first assertion is just incorrect. There are 2.1 *quadrillion* Satoshis in total, most already mined. All the wealth in the world currently totals less than half that many units as measured in dollars, including all real estate, stock and bonds, commodities and other goods that will never be used as "money". Remember also that the bitcoin community is able to make improvements to the platform via consensus for the good of all. It takes a huge effort, but this is no doubt what would happen if the cryptography were somehow cracked or we ran out of units and needed to add more ooos. Note that dividing units further does not inflate the supply of total units; this is not monetary inflation; it is simply adding a new denomination.

Secondly, the ten-minute block time and the block size are intentionally limiting for good reason. For purely physical reasons, a truly decentralized system will need to allow some time for nodes to disseminate, process and verify transactions.

Bitcoin transactions are considered 'finally settled' when enough subsequent blocks (usually six) have been attached to the blockchain such that undoing the work is prohibitively costly to an attacker. The emphases in a value transfer system are security and accuracy over speed. Remember that the **final settlement** in the traditional system takes somewhere around **six to eight weeks** after audits are closed. *Bitcoin is not slower than traditional transactions, even Visa.* Those who trumpet transaction speed conveniently ignore settlement speeds.

Even the Altcoins that mutated Bitcoin into 'faster' transaction speed do not reach final settlement as fast as Bitcoin's approximate one hour because volume in the Altworld is so much smaller that confirmations take longer to accumulate.

Satoshi knew that, like the protocols of the internet, many layers could be built upon the Bitcoin rails he was creating. The best analogy I have heard about how to imagine a second layer on bitcoin is as follows:

Illustration 49: Second Layer Bar Tab

When the beer you have ordered arrives, the bartender will usually ask you one question: would you like to close out or start a tab?

If you close out, your relationship with the bartender ends with the recording of the transaction on the payment layer, which is final as you take custody of the beer and your payment card is returned.

If you 'start a tab', you have effectively shifted your relationship with the bartender to a 'second layer'. You have opened a more casual two-party relationship with the barkeep whereby you can signal your needs, and they will be tracked "off-chain" on a tally.

This second layer is much more convenient, faster and cheaper for both you and the bartender, as there are fees for every base layer transaction and a drag on the barman's time. A second layer does not tie up time or space on the main chain because either your identity, creditworthiness, or collateral has satisfied the bartender enough to provide you drinks before settlement. At your departure, you 'settle up' your tab with a single transaction that reflects all your business with the barkeep and ends your relationship. Any hold the barman had on your verification, such as your license or credit card, is returned. I agree that at Cheers, Coach might have let you slide a week—but this is the real world.

As we have discussed with tokens, we already employ second-layer technology in many places. Venmo is the TradFi version of what the Lightning Network does for Bitcoin. They are both faster, peer-to-peer transactions on a second layer. Second layers can use different ways to verify the 'trustworthiness' metric for transactions with strangers; commonly the asset tokens are 'siloed' for the verification or transaction logistics elements and then released.

A last thought about scaling. We often describe scaling in terms of the number of users or entities of geographic spread. There is also a scale that represents unit levels at which various entities tend to do business, or the distribution of money. As denominated in US dollars, Nation-States deal with currency on a sovereign level and their central banks handle figures into the trillions. Banks and large financial institutions deal with balance sheets often in the billions of dollars. Larger businesses of all sorts tend to deal in the millions of dollars, small businesses deal in thousands, and the average person worldwide is really only dealing with amounts below $1000 the vast majority of time.

Obviously, if only one unit of measurement exists at a very wide range of distribution, there need to be many units available to ensure liquidity at both the individual and the Sovereign state levels. This is not an issue for unlimited fiat currencies, so we are not exactly sure how we will react to a strictly fixed unit of measure. Scaling for these cohorts may end up happening on different layers, or bitcoin's price increases might make a whole coin quite valuable...or both. I do foresee that we will have nicknames for all the denominations between a whole bitcoin and a Satoshi, as they will be used by these different spending cohorts.

"With e-currency based on cryptographic proof, without the need to trust a third-party middleman, money can be secure and transactions effortless. "

– Satoshi Nakamoto

Satoshi created a protocol that has kept to his vision. It has proven itself in world programming circles and in world markets. The question is when we will be able to create and properly regulate further iterations of the technology—if different versions are indeed necessary. Maybe we will create many different layers that parallel Bitcoin and use it as a base layer.

However, it happens that Bitcoin is eventually used, I do not believe it will happen suddenly. As already has been the case, Bitcoin network use and token adoption will rise with education and understanding. Because Bitcoin is permissionless, governments cannot effectively say you cannot use it. They can say that you must or can use bitcoin, which is another asymmetric advantage to the decentralized platform.

Even in modern times, the conversion of one money system to another usually takes a while to effect. Any money is not worth much until enough trading partners agree to honor it, and we are fond of entrenching our systems and building in conservative power-preserving structures to benefit the incumbents. It may very well take even longer for an unrepresented, truly decentralized system to bootstrap its way up to the sovereign/central bank level.

Impressively, though, bitcoin's market cap is larger than the largest US Commercial or Investment Bank, so it does seem to be headed toward the next level. Using a market cap measure, by simply multiplying price by number of coins, one fails to account for all the investment in mining, custody solutions, and all other platform, exchange and banking value being added every day to the technology.

In 2021, El Salvador became the first nation to adopt it as a national legal tender.

"It is fascinating that in discussions about what is a 'safe asset,' bitcoin is increasingly entering the conversation, and that's a significant change"

– Mohamed El-Erian

The other exciting element to add to adoption is the growing awareness that Bitcoin's transparency not only is a foil *against* criminal behavior, but that it also gives us numbers nerds a whole new data set with which to study a market. Unlike in the world of traditional finance, the details of each transaction on a blockchain are preserved and available to all.

This means that we have a new universe of data available to study behavior and economics. Bitcoin analytics can tell us things that traditional markets cannot: the number of wallets, the number of wallets in profit, the specific cohorts of buyers or sellers, etc. Instead of using a rough market cap measurement, we can use a realized price basis, aggregating the prices that were actually last paid for any given Satoshi. Our ability to parse all Bitcoin transactional data is revolutionary compared to the opacity of traditional asset markets. And I can keep Bitcoin's entire fifteen-year transactional history on an $80 external drive on my PC at home. Wow.

It seems to me foolish to bet against human ingenuity and a permissionless and fair network that might be the ultimate vehicle for free expression.

The future is *digihuman.*

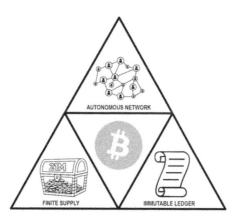

"Writing a description for this thing for general audiences is bloody hard. There's nothing to relate it to."

– Satoshi Nakamoto

PARTIAL LIST OF IDEAS INCORPORATED INTO BITCOIN

- Consensus Protocols
- Cooperative Security
- Cooperative Verification
- Decentralization/Decentralized and Autonomous Network
- Difficulty Adjustment
- Disinflationary Supply Schedule
- Finite Supply/Inelastic Curve
- Immutable Ledger
- Medium of Exchange
- Money
- Node to node transactions
- Offsetting public liabilities (landfills)
- Permissionlessness
- Proof of Work

- Real World, Physical cost to attack
- Store of Value
- Token Types—Utility, Identity, Ownership, Permission, Value
- Unit of Account

NOTES

INTRODUCTION

1. I am following the convention of using a capitalized "Bitcoin" to refer to the network and technology and bitcoin with a small "b" to refer specifically to the tokens of value.
2. In its very early years, two bugs were identified and fixed by developers. These resulted in system downtime of about 5 hours. That was over ten years ago.

THE HALVENINGS

1. Some people call these the "halvings" some call them "halvenings"...whatever.